APICULTURAL LITERATURE PUBLISHED IN

CANADA AND THE UNITED STATES

T. S. K. Johansson
Queens College

and

M. P. Johansson
Queensborough Community College

of

The City University of New York

"... I believe we all are conceited about what we know
about the busy little bees. After reading the modern
works, and then the ancient, it sort of knocks the props
out. We ought to give those old fellows some honor. I
feel that we have borrowed it all, or most all, from
them."

Chas. E. Kemp (1901)

1972

ISBN: 978-1-908904-17-1

Published in 2012 by

Northern Bee Books
Scout Bottom Farm
Mytholmroyd
Hebden Bridge HX7 5JS (UK)

Printed by Lightning Source, UK

Cover image Kim Flottum

PREFACE

This bibliography is based primarily on the contents of ten major library collections. The compilation was started after the late C.L. Farrar of the U.S. Department of Agriculture invited us, in 1959, to collate the Miller Memorial Collection which had been put into storage. We are grateful for the courtesies extended to us by E.M. Pittenger at the Steenboch Library, J.S. Merrill at the Bee Culture Branch of the National Agricultural Library, R.A. Grout at Dadant & Sons, Inc., and the Library of Congress; for copies of shelf lists from the University of California Library at Davis and the Entomology Library of the University of Minnesota; for a printout of the holdings in the University of Guelph Library; and for bibliographic listings from the Bee Research Association.

We would appreciate hearing of omissions (and errors) which we will compile and send to all who make such contributions. Others can secure a copy by request with enclosure of a stamped self-addressed envelope (UNESCO Postal Coupon from foreign addresses). Please send full particulars and indicate whether the item can be loaned or pertinent pages copied for verification. Relevant bibliographical, biographical, and historical notes would also be appreciated. Many author's birth and death dates are missing (see Author Index).

There are other valuable public or private collections which we have not seen and we would welcome copies or references to lists of such collections as well as lists of duplicates or collections offered for sale.

T.S.K.J
M.P.J.

Please send all correspondence to:

M.P. Johansson, Department of Biology
Queensborough Community College
Bayside, New York 11364
U.S.A.

CONTENTS

BIBLIOGRAPHIC ESSAY

If there is one subject in natural history that generates more
interest than almost any other, it is the behavior of social insects
and particularly the honey bee. This interest is recorded in stone-
age drawings, Pliny's account of hives with thin horn walls through
which guests of a Roman consul could observe the wonders of a bee colony,
and the first anatomical drawings using a microscope. Today, an observation
hive on display at a fair, museum, store window, or class room continues
to attract attention.

At the first American Beekeepers' Convention in Cleveland, Ohio
(March 15, 1860), L.L. Langstroth, the "father of American beekeeping",
read excerpts from Columella (60 A.D.) and Charles Butler (1634) to
substantiate points of argument. When Langstroth sent A.I. Root a copy
of Thomas Wildman's Treatise on Bees (1768) and John Keys' Antient Beemaster's Farewell (1796), Mr. Root was surprised to find material that
he thought had been discovered only recently - the use of contrasting
colors on hives to minimize drifting: "Is it really true that there
is nothing new under the sun?"

Inevitably the literature on apiculture in the United States and
Canada has reflected the technology and economics of the industry as
well as political and agricultural developments in these countries.
Hence a brief historical correlation may be helpful.

1639-1821. There is strong presumptive evidence that the honey
bee (Apis mellifera L.) was not native to the New World but was brought
from England, Holland, and Spain by the early settlers [See Barton
(51); Belknap (54)]. Certainly, beekeeping was an integral part of
colonial agriculture supplying a sweetening agent and wax for candles
particularly valued by the Catholic settlers. Swarms escaping into the
wilderness surrounding the settlements were abundant enough for the
Indians to coin the label "white man's fly".

A few seventeenth century printing presses were brought to English
America (first to Cambridge, Mass. in 1639) primarily as tools for the
clergy and government. Books were largely imported but the colonial
printers gradually combined the roles of printer, publisher, newspaper
editor, compiler, and author. Governmental control of the press culminated
in the Townshend Act of 1767 which included a tax on imported paper as
well as tea, and the colonial printers added their sparks to the 1776
Revolution. The first book in the New World devoted exclusively to bees
was published in 1792 by Isaiah Thomas, one of the most prominent of the
eighteenth century printers with 16 presses and bookstores in the major
cities. Relatively few books were published before 1821 when mechanical
improvements in printing made mass production possible, and only a
handful of these contained material on apiculture:

```
1792 Thomas, I.  (publ) [Taken from D. Wildman]
1792 Belknap, J.
1793 Barton, B.S.
1795 Day, T.  [Juv]
1803 Anon  A short history of bees.  [Repr of English book]
1804 Miner, T.
1807 Souder, D.  [Taken from J.L. Christ]
1813 Doddridge, J.
```

1819 Wiestling, J.S. [In German]
1819 Butler, F.

1821-1852. Before 1853, there was little commercial beekeeping.
The majority of the population lived on farms and almost every household
had a shed in which stood tall box-hives, straw skeps, or an occasional
section of a bee tree. J.E. Crane described the methods used in harvesting
honey. In October, when brood rearing had ceased, sulphur "candles"
were made and burned in a pit to suffocate the bees in the lightest hives
[Glean. Bee Cult. 59(7): 427-29 (1931)]. New white combs were put aside
for table use, and the darker brood combs broken up and drained to produce
"strained honey". Towards the end of this period the practice of placing
a "cap" over an auger hole in the top of the hive enabled some surplus
box-honey to be marketed for cash.

Of the 30 volumes published during this time, 19 were written for
the apparent purpose of promoting "patent" hives and devices invented
by shrewd Yankees purporting to solve difficult bee management problems
such as swarming, increase, wax moths, etc. Even Langstroth's patent
application in 1852, for the prototype of the modern hive, included a
moth trap in tune with the "humbuggery" of the times. Between 1790 and
1873 nearly 600 beekeeping devices were patented.

Two British books reprinted in America are worthy of mention: the
1843 reprint of E. Bevan's second edition (1838) of The Honey Bee containing
the crucial observations and experiments of R. Golding which provided the
model for Langstroth's movable frames [See Bee World 48(4): 133-43 (1967)],
and the 1846 reprint of W. Kirby and W. Spence's sixth edition of An
Introduction to Entomology which included what was known of bees from
classic and contemporary writers, and referred the reader to Dr. Bevan
for "much valuable information on the economy of bees". The latter book
was printed in Greenfield, Mass. while Langstroth was a resident there
(1840-1848) by the same shop that printed Langstroth's book in 1853.

Most farmers knew little of what went on inside the hive and depended
upon natural swarms for increase. Dr. J.V.C. Smith, whose book was the
first one read by Langstroth, even doubted the existence of the queen bee.
Agricultural journals were numerous, and beekeeping articles and letters
with frequent quotations and references to Bevan, and Kirby and Spence
were the major source of information. The American Farmer was the first
"farm paper" (1819-1833). The Cultivator begun in 1831 claimed to "improve
the soil and the mind" and is still fascinating reading - including the
advertisements.

1853-1867. Box-honey. The 1853 publication of books by L.L. Langstroth
and M. Quinby is a bench mark for the beginning of modern apiculture.
Quinby was a successful commercial beekeeper, and his book did much to
stimulate the production of box-honey - glass-sided wooden boxes containing
2 to 16 pounds of honey. Langstroth's The Hive and the Honeybee, was the
best source of information about behavior and management of bees, and
went through three editions and many reprintings until 1889 when it was
revised and enlarged by Chas. Dadant & Son, manufacturers of bee supplies.
The first edition had an advertisement for Langstroth's patent but no
specifications or illustrations of his hive. These were added in the
second edition (1857).

A crucial stimulus to the development of beekeeping was the "fever" attending introduction of Italian bees into the U.S. in 1860 [Glean. Bee Cult. 98(12): 720-24, 755 (1970)]. The prospect of riches from producing queens for requeening and sale necessitated using movable frame hives such as Langstroth's.

1867-1876. Extracted honey I. The invention of the centrifugal extractor in 1865, for spinning out honey from movable frames, made it possible to return the wax combs for refilling. Unfortunately, marketing liquid honey in bottles gave rise to adulteration with cheaper glucose. The promotion of patent hives and extractors made little impact on the beekeeping of the majority of subsistence farmers, and box hives were common until the 1880's. In some sections of the U.S., "modern" hives were not prevalent until the widespread prosperity following World War II.

1876-1906. Comb-honey. Two inventions, the one-piece, one-pound section box in 1873, and the machinery for making sheets of thin comb foundation (midrib) embossed with the bases of cells in 1876, made section comb a commercial possibility replacing the larger box-honey. Management of the colonies became more exacting since it was necessary to crowd the bees to force them into the sections and advocacy of a small brood chamber exacerbated the swarming problem. Successful producers of comb honey such as C.C. Miller were keen observers of bee behavior and their writings are still in demand. Beekeeping journals during this "golden age" (through the 1920's) document the original observations and ideas on which present practices are based.

1906-____. Extracted honey II. The restoration of confidence in the purity of liquid honey following passage of the Federal Pure Food and Drug Law in 1906, the rapid spread of legumes over the plains, and the sugar scarcity of World War I induced expansion of beekeeping to its present level. Since World War II, beekeeping has been vitally affected by agricultural practices designed to meet increased costs with greater efficiency - removing fence rows and brushy corners, harvesting forage legumes as they bloom, introducing new varieties of legumes which are dependent on good weather for nectar production, increasing the use of pesticides and herbicides, combining small, diversified farms into single crop, mechanized agricultural units, and conversion of agricultural lands to housing or industrial uses.

The number of commercial beekeepers has never been large but there are many persons interested in keeping a few hives as a sideline or hobby. Among them are serious bibliophiles with excellent collections and extensive knowledge of the literature. In the hope that this bibliography will have general usefulness to apiculturists, we have refrained from imposing our own limits and have included all the items which came to hand.

ABBREVIATIONS

Abbreviations conform to those listed in the Style Manual for
Biological Journals [A.I.B.S. (1964 2nd ed)]. Terms not listed in the
above reference were abbreviated according to the GPO Style Manual
[U.S. Government Printing Office (1959) or Webster's New International
Dictionary (1958 2nd ed).

Abr	Abridged	Front	Frontispiece
Abs	Abstract	Gen	General
Acad	Academy	Glean	Gleanings
Admin	Administration	Glos	Glossary
Adv	Advertisement(s)	Hist	Histor(ical)(y)
Advis	Advisory	Hort	Horticultur(al)(e)
Agr	Agricultur(al)(e)(ist)	Illus	Illustrations, plates,
Amer	America(n)		figures, tables, frontis-
Anon	Anonymous		piece
An	Another	Inc	Incorporated
Apicult	Apicultur(al)(e)	Inform	Information(al)
Append	Appendix	Inst	Institut(e)(ion)
ARS	Agricultural Research	Int	International
	Service	J	Journal
Ass	Association	Juv	Juvenile literature
Bd	Board	Leafl	Leaflet
Bibliogr	Bibliograph(ical)(y)	Libr	Library
Bro	Brother(s)	Man	Manual
Bull	Bulletin	Mfg	Manufacturing
Bur	Bureau	Mimeo	Mimeographed
C	Copyright	Misc	Miscellaneous
Cat	Catalog	Mktg	Market(ing)(s)
Can	Canad(a)(ian)	Monogr	Monograph
Centen	Centennial	Mus	Museum
Chem	Chemi(cal)(stry)	Nat	National
Circ	Circular	N D	No date
Co	Company	No	Number
Col	Column(s)	N S	New series
Coll	College	Offic	Official
Collect	Collection(s)	P	Page(s)
Comm	Commi(ssion)(ttee)	Pam	Pamphlet
Comp	Compile(d)(r)	Pat	Patent(ed)
Cult	Culture	Polytech	Polytechnic(al)
Dep	Department	Prelim	Preliminary
Div	Division	Print	Printer
Doc	Document	Prod	Produc(er)(tion)
Econ	Econom(ic)(y)	Pseud	Pseudonym
Ed	Edit(ed)(ion(s))(or)	Pub	Publication(s)
Educ	Education(al)	Publ	Publish(ed)(er)
Enl	Enlarged	Rep	Report
Entomol	Entomolog(ical)(ist)(y)	Repr	Reprint
Exp	Experimental	Res	Research
Ext	Extention	Rev	Revi(ew(ed))(s(ed)(ion))
Farm	Farm(er(s))(ing)	Sci	Scien(ce)(tific)
Fed	Federal	Sect	Section
Fedn	Federation	Ser	Series
Found	Foundation	Serv	Service

```
Soc       Society
Spec      Special(ist)
Sta       Station
Suppl     Supplement
Tech      Technical
Trans     Transaction(s)
Transl    Translat(ed)(ion)(or)
Univ      University
Vol       Volume(s)
```

BOOKS AND PAMPHLETS

The following pages list publications by private individuals, bee-keeping organizations and business enterprises. Federal, state, dominion, and provincial publications are listed in separate sections. The intent was to include writings directly concerned with honeybee biology and behavior, and the technology of beekeeping and products of the hive. Such topics as pollination per se which would make a considerable bibliography of its own have been omitted. A few early agricultural and general entomology books which consider bees or beekeeping in detail have been included. Fiction, children's books [See British Bee J. 82: 846 (1954) for juvenile books in the Bee Research Association Library], or honey recipe books are so indicated.

Publications are listed alphabetically by author or organization if known. Where the above information could not be determined, publications are listed as anonymous and arranged alphabetically by title. Works by the same author are arranged chronologically. Information enclosed in square brackets has been determined [--] or implied [--?] from sources other than the publication itself. Where the publisher and place of publication is not indicated in the book, the name and location of the printer has been used (if known). States or provinces have been omitted for the following cities:

Baltimore, Md.	Medina, Ohio	St. Louis, Mo.
Boston, Mass.	New York, N.Y.	San Francisco, Calif.
Chicago, Ill.	Ottawa, Ont.	Toronto, Ont.
Cincinnati, Ohio	Philadelphia, Pa.	
Hamilton, Ill.	Pittsburgh, Pa.	

Books first published elsewhere have been included with a notation of the place and date of original publication but a complete reference to all editions outside the U.S. and Canada is not included. This also applies to translations.

Holdings at the libraries listed below are given for items published before 1940 and were checked at the libraries with the exception of the University of Minnesota and the Bee Research Association. Where no library is indicated, the information was obtained from private collections, bibliographies, book catalogs, or advertisements and reviews in the bee journals.

Abbreviations for Libraries

B - Bee Research Association, Hill House, Chalfont St. Peter, Gerrards Cross, Bucks., England
C - Mann Library, Cornell University, Ithaca, N.Y. 14850
E - University of California Library, Davis, Calif. 95616
F - New York Public Library, 5th Ave., New York, N.Y. 10036
I - Dadant & Sons Inc., Hamilton, Ill. 62341
L - Entomology Library, University of Minnesota, Minneapolis, Minn. 55455
M - Steenbock Library, University of Wisconsin, Madison, Wis. 53706
T - University of Guelph Library, Guelph, Ont.
W - Library of Congress, Washington, D.C. 20540
X - National Agricultural Library, Washington, D.C. 20250; Bee Culture Branch, Beltsville, Md. 20705

The books listed can be used at the libraries or borrowed through interlibrary loan arranged at a local library, with the exception of rare books or those in delicate condition. The definition of "rare" varies. In one library all books published before 1900 are under lock and key irrespective of the ease of replacing them. One out-of-print book published in 1951 is harder to replace than some published in 1875. Xerox copies may be obtained from some university libraries for a charge of 10 cents per sheet and $1 to $1.50 for handling (1970 prices).

Dadant & Sons Inc., manufacturers of beekeeping supplies, and publishers of the Amer. Bee J. and books on beekeeping, do not have facilities for either interlibrary loans or copying, but items not available elsewhere can be used at their plant for research purposes if arrangements are made in advance.

"My jokes need labeling sometimes. Speaking of old bee-books, I said none were published in America before 1492. Now comes a grave statement in a foreign bee-journal, that the first bee-book in America was printed in 1492."

C.C. Miller (1894)

1   ANON   nd   The "glassel" hive stand. 1p illus   C
2   ANON   1859   Honey blossoms for little bees. N.Y., M.W. Dodd
vi,8-236p illus juv
2a   ANON   1933   The importance of honey for health. N.Y., John
G. Paton Co 28p recipes   F [uncataloged]
3   ANON   1830   The natural history of insects. First series.
N.Y., J. & J. Harper 292p illus (The hive bee p.25-81)   M
    1831 & 1832   I;   1843 Harper & Bro   I M;   1856   M
4   ANON   1803   A short history of bees. In two parts. Philadelphia,
J. Johnson xiv,50,2p   M
    Publ London 1800 133p illus. Many ed   B
    [See Amer. Bee J. 73(9): 367 (1933)]
5   ANON   nd   Some characteristics of the bee-keepers' review.
Flint, Mich. 20p illus   I
6   ANON   [1904?]   The Wiley "Honey lie". A sciencific pleasantry.
Documents in evidence. 23p   E M X
    [See Glean. Bee Cult. 99(4): 132-33, 150 (1971)]
7   ABBOTT, E.T.   1886   The busy bee. Or beekeeping in a nutshell. St. Joseph, Mo., C.P. Kingsbury 38p illus   C W
    St. Joseph Apiary price list p.29-37
8   ABBOTT, M.   1965   Me and the bee; the battles of a bumbling
beekeeper. Annapolis, Md., Hammond-Harwood 189p illus
9   ADAIR, D.L.   1867   A new system of bee-keeping; adapted to the
habits and characteristics of the honey-bee: with descriptions of, and
directions for managing bees in the section bee-hive. Embracing also
improved methods of artificial swarming..., whereby the business of beekeeping is rendered more profitable and pleasant. Cincinnati, R. Clarke
& Co 74p illus   W X
    Contains adv for Adair's section bee-hive
10   ADAIR, D.L. (ed)   1869   Annals of bee culture for 1869, a
bee-keepers' year book, with communications from the best American apiarians
and naturalists. Louisville, Ky., Hull & Bro, Print 57p illus   I M T X
    1870 C.Y. Duncan 64p illus   I M X
    1872 John P. Morton & Co 64p   M W X
    1872 The Author   F C
    1869-72 Includes 1869, '70, '72 ed & Progressive bee culture [12]   C M
11   ADAIR, D.L.   [1870?] 2nd ed   Outlines of bee culture and
descriptive catalogue. [Cincinnati], Robert Clarke & Co 22p illus   I
12   ADAIR, D.L.   1872   Progressive bee culture or Apine instincts
and labors defined, illustrated and systematized, upon a new theory.
Cincinnati, The Author 24p illus   C I M W X
    Contains price list of equipment including Adair's section bee-hive
    and meliput honey extractor. See 10
13   ADAMS, C.F.   1923   The Adams double walled convertible beehive.
Spencer, Mass. 6p illus   C
    Includes photograph of his bee shed described in Amer. Bee J. 62(12):
    560-61 (1922)
14   ADAMS, T.J.   1904   Honey fairies in the sunny hills of fair
Columbia; N.B.K.A. souvenir. With extracts from Campers in Fairy Land.
St. Louis 22p illus   E M
    Poetry for World's Fair, Louisiana Purchase Exposition; Nat. Bee-
    keepers' Ass. Convention
15   AFFLECK, T.   1841   Bee-breeding in the West. Cincinnati,
E. Lucas 70p illus   C I M W X
    Ed (with E.J. Hopper) Western Farmer & Gardner and developed subtended
    or mutliple box hive

16   ALDRICH, C.C.   1890   Hand book of bee culture. Designed
for the use of the beginner in beekeeping in the Northwest. Morristown,
Minn., Press of Hollister Bro   42p   W
17   ALEXANDER, E.W.   1909   Alexander's writings on practical
bee culture. Ed & comp by H.H. Root. Medina, A.I. Root Co 96p illus F M
    1910 2nd ed   C L T X;   1910 3rd ed 98p   B C E I L M T
    A renowned bee-master of Delanson, Schenactady County, N.Y.
    Selections from articles appearing in Glean. Bee Cult. between Nov. 1,
    1904 & March 15, 1908
18   ALLEN, T.R.   1846   Allen's bee cultivator, embracing the
natural history, physiology and management of the honey bee. Syracuse,
[N.Y.], S.F. Smith & Co 58p illus   C X
19   ALLEY, H.   1883   The bee-keeper's handy book: or Twenty-two
years' experience in queen-rearing, containing the only scientific and
practical method of rearing queen bees, and the latest and best methods
for the general management of the apiary. Contains essays by Geo. W.
HOUSE and Silas M. LOCKE respectively: (1) Management of the apiary;
or, the production and marketing of honey and (2) The new races. Wenham,
Mass., The Author c1882 xv,184p illus [Alfred Neighbor's copy (M)
has different adv]   B C E F I M T W X
    1885 3rd ed   xix, 269p   C I L M T X
    Preface mentions that 1st ed of 2,000 was disposed of in 18 months
    Adv include Alley's Bay State Apiary (established 1857) "devoted wholly
    to queen rearing"
20   ALLEY, H. (comp)   1889   The national beekeepers' directory.
Containing a classified list of the beekeepers of the United States and
Canada; with essays and hints regarding the successful management of the
apiary. Salem, Mass., Salem Press Publ & Print Co (c George A. Bates)
139p illus adv   C M X
    Lists less than 3,000 persons responding to 10,000 requests and keeping
    an average of 20 colonies averaging 25 lb of honey per season. History
    of bee periodicals in America p. 132-34
21   ALLEY, H.   1891   Thirty years among the bees. The results
of a quarter-century experience in rearing queen-bees, giving the practical,
every-day work of the apiary. Salem, Mass., The Author 72p illus   C X
    1893 88p   C E M
22   ALLEY, H.   1898   Successful methods for rearing queen bees
as practiced over thirty years. Wenham, Mass., George E. Frost 18p
illus   M
    1899 2nd ed Beverly, [Mass.], Frost & Wood 20p   X
    Includes Alley's price list
23   ALLEY, H.   1903   Improved queen-rearing or how to rear large,
prolific, long-lived queen bees. The result of nearly half a century's
experience in rearing queen bees, giving the practical, every-day work of
the queen-rearing apiary. Beverly, Mass., The Author 55p illus   I M X
    Alley offers to sell equipment
24   AMER. HONEY INST.   nd   100 Honey helpings. Madison, Wis.
31p illus recipes
25   AMER. HONEY INST.   nd   Using honey. Madison, Wis. [4p+table]
recipes
26   AMER. HONEY INST.   1941   Old favorite honey recipes. Madison,
Wis. 40p illus
    1945 rev enl 53p
27   AMER. HONEY INST.   [1944]   Honey. Madison, Wis. [10p] recipes
28   AMER. HONEY INST.   1947   New favorite honey recipes. Madison,
Wis. 56p illus

29   AMER. HONEY INST.   [1949] rev enl   Honey specialties for bakers.
Madison, Wis. 49p
30   AMER. HONEY INST.   1955   Cookies made with honey. Madison,
Wis. 31p illus
31   AMER. HONEY INST.   [1959]   Honey recipes. Large quantity.
Madison, Wis. 61p
     Mimeo ed 27p;   An mimeo ed 28p
32   AMER. HONEY INST.   1962   Made with honey. Madison, Wis.
15p recipes
33   AMER. HONEY INST.   [1963]   Honey holidays. Madison, Wis.
17p recipes
34   AMER. HONEY INST.   [1963]   Lemons and honey. Madison, Wis.
11p illus recipes
35   AMER. HONEY INST.   [1963]   Milk and honey treasures. Madison,
Wis. 11p illus recipes
36   AMER. HONEY PROD. LEAGUE   1924   A treatise on the law pertaining
to the honeybee by the legal department of the American Honey Producers'
League. Madison, Wis. In cooperation with Amer. Bee J. & Gleam. Bee
Cult. 88p   B C E I M T W X
37   ARNOLD, D.J.   1874   Practical hints on bee culture. Brownville,
Nebr., Advertiser Book & Job Print 52p illus   W X
     Offered territories for Arnold's improved movable comb hive
38   ATCHLEY, J.   nd   Lessons in profitable bee-keeping. Also
bees, queens and bee-keepers' supplies. Beeville, Tex., Southland Queen
Print iv,2,5p,price list xii,15 lessons,28p illus   T
     [1897] 36p E;   1898 8th ed Catalogue and price list of the Jenny
     Atchley Co 16 lessons 45p   B
39   ATHERTON, G.   1941   Do bees see in the dark? St. Paul, Minn.,
The Author [11p]
40   ATKINS, E.W. & HAWKINS, K.   1924   How to succeed with bees...
[Watertown, Wis., G.B. Lewis Co] 96p illus   B E F I L M T W X
     [1925] 2nd ed   B C I;   [1926] 3rd ed   B C E;   [19277] 4th ed
T;   [1931] 6th ed   M X;   [1936] 8th ed   I;   [1937] 9th ed
C I M X;   [1938] 10th ed   C X;   12th ed   C I;   [1942] 13th
ed;   [1942] 14th ed;   [1942] 15th ed;   16th ed;   [1944]
17th ed;   [1945] 18th ed;   [1946] 19th ed
     Each chapter has an outline and a set of review questions.
     Authors were associated with G.B. Lewis Co as sales promotion & general
     sales manager respectively
41   AVEBURY, J. LUBBOCK   1882   Ants, bees, and wasps. A record
of observations on the habits of the social hymenoptera. N.Y., D. Appleton
& Co xix,448p illus bibliogr   F W
     1883   F;   1884   E T X;   1888   W;   1892   F;   1893;
     1894   E;   1895   E;   1899   T;   1904   F;   1906   F;
     1911   W;   1915   X;   1917   T
     1929 ed & annotated by J.G. Myers xix,377p   E M W X
     1971 Lexington, Mass., Gregg Intl Publ North Amer
42   B., de   1863   The honey-makers. Boston, Amer Tract Soc
110p illus   I M T
43   BAIN, F.W.   1924   A syrup of the bees. (Transl from Hindu)
xix,144p   I
44   BALDWIN, E.G.   1921   Beekeeping in Florida. Medina, A.I. Root
Co 23p (C has 16p ed)   M T
     Author former Fed Ext Spec in beekeeping
45   BALDWIN, L.   1945   Beekeeping. Occupational Abs 79. N.Y.
Occupational Index, N.Y. Univ 6p

46   BALLANTINE, W.   1877   Beeology; or, A brief, practical
treatise on the management and culture of bees, in connection with the
twin bee-hive. Zanesville, Ohio, Courier Print 28,2p illus   E
47   BALLANTINE, W.   1884   A practical treatise on bee culture;
for profit and pleasure. Sago, Ohio, Bloomfield Print Office xii,160p
illus   C E F I M X
48   BARNARD, H.E. & FISCHER, M.D.   [1929]   Honey in the bakeshop.
Indianapolis, Ind., Amer Honey Inst 29p illus   M T
49   BARNES, J. & W.   nd   First lessons in bee culture. St. Louis,
T.S. Bowman 44p   C
50   [BARTMAN, M.]   1940   The story of bees, compiled by workers
of the writers' program of the work projects administration in the common-
wealth of Pennsylvania. (Children's Science Ser 4) Chicago, Albert
Whitman & Co 45p illus juv
51   BARTON, B.S.   1793   An inquiry into the question, whether the
Apis mellifera, or true honey bee, is a native of America. Trans. Amer.
Philosophical Soc. 3: 241-61
52   BECK, B.F.   1935   Bee venom therapy. Bee venom, its nature,
and its effect on arthritic and rheumatoid conditions. N.Y. & London,
D. Appleton-Century Co xii,238p bibliogr   B C I T W X
53   BECK, B.F.   1938   Honey and health. N.Y., Robert M. McBride
& Co xiv,272p illus bibliogr   B C E F I L T W X
    1944 rev Beck, B.F. & SMEDLEY, D. Honey and your health. 246p
54   BELKNAP, J.   1792   A discourse, intended to commemorate the
discovery of America by Christopher Columbus; delivered at the request of
the Historical Society of Massachusetts, on the 23d day of October, 1792,
being the completion of the third century since that memorable event. To
which are added four dissertations, connected with various parts of the
discourse... Boston, Printed at the Apollo Press by Belknap & Hall
132,2p adv   F
    Diss. III On the question, whether the honey-bee is a native of
    America? [p. 117-24]
55   BENEDICT, A.   1873   Honey bee. Bennington, Ohio
56   BENTON, F.   1895   Apis dorsata. Giant bees in India.
Jamestown, N.Y., W.T. Falconer Mfg Co [4p]   C
57   BENTON, F.   1895   The bees for the harvest. Wash., D.C.,
Amer Farm Print 5p illus   C
    Read before the Ind. Bee-Keepers' Ass., Jan. 10, 1895
58   BERLEPSCH, A.   1877   The honey bee. The Dzierzon theory;
being a full elucidation of scientific bee-keeping, by the Baron of Berlepsch.
Transl S. Wagner. Chicago, Office of the Amer. Bee J. 5-50p   M
    1882 48p   C M;   1884 Medina, A.I. Root   T X;   1902   C I
    Publ in 1st vol of Amer. Bee J. (1861)
59   BERLINER, J.J. & STAFF   nd   A Berliner research report on
honey. N.Y. 59,1p ditto   I
60   BESSONET, E.C.   nd   Modern queen rearing and package bee
production. Hapeville, Ga., Hale Publ Co
    Adv on wrapper of Latham 288 but never published
61   BEVAN, E.   1843   The honey bee; its natural history, physiology
and management. Philadelphia, Carey & Hart viii,128p illus   B C F M X
    Publ England 1827, other ed
62   BIGELOW, E.F.   [1905]   The educational bee-hive. Medina,
A.I. Root 15p illus   C
63   BILL, A.C.   1926   The message of the bee. Wash., D.C.,
A.A. Beauchamp 77p   C W
    Analogies to life in the hive (Christian Science)

64   BONSELS, W.   1922   The adventures of Maya the bee. Transl
A.S. Seltzer. N.Y., Thomas Seltzer 224p illus   B C F I L
     Publ England, various dates
65   BOSWELL, P.   1842   Bees, pigeons, rabbits, and the canary
bird, familiarly described. N.Y., Wiley & Putnam 160p [misnumbered
164] illus   I M W X
66   BRADSHAW, D.B.   [1960]   Why federal marketing agreements
and orders cannot be successfully applied to the honey industry. [Idaho]
11p
67   BRITE, C & KROCHMAL, A.   1969   Cooking with honey. Jewish
honey recipes. Berea, Ky., The Authors 29p
68   BROADMAN, J.   1962   Bee venom; the natural curative for
arthritis and rheumatism. N.Y., G.P. Putnam's Sons 220p bibliogr
69   BROWN, A.C. (comp)   1957   Granny's honey and beeswax pre-
scriptions. Rutland, Vt., Tuttle Publ Co 64p
70   BROWN, H.   1923   A bee melody. N.Y. & London, Andrew Melrose
vi,274p   .   B L M T
71   BROWN, J.P.H.   1898   Bee-keeping for beginners. A practical
and condensed treatise on the honey-bee. Giving the best modes of management
in order to secure the most profit. Augusta, Ga., Richards & Shaver, Print
iv,110p illus   C E I L M W
     Hardcover and "paperback"
72   BROWN, W.   1824   Compendium of agriculture, farmer's guide
in the lost essential parts of husbandry and gardening: with the aid
and inspection of Solomon Brown. Providence, [R.I.] 228p   M
73   BURGESS, T.W.   1927   Milk and honey. Racine, Wis., Whitman
Publ Co 22p illus juv
74   BURROUGHS, J.   1879   Locusts and wild honey. Cambridge,
[Mass.], Riverside Press 253p
     The pastoral bees p. 7-34
     1901 229p   M;   1905 235p   I
75   BURROUGHS, J.   1887   Birds and bees. Boston, Houghton,
Mifflin & Co c1879,'81,'86 Riverside Literary Ser 28 88p
     Many ed
76   BURT, M.E.   1889   Bees; a study from Virgil, revised and
·adapted from Davidson's translation for seventh grade. Chicago, S.R.
Winchell & Co 15p   W
77   BUSCH, W.   1873   Buzz a buzz. Transl Hezekiah Watkins. N.Y.,
Henry Holt & Co 80p illus juv   I;   Publ England 1872   M
78   BUTLER, F.   1819   The farmer's manual; being a plain practical
treatise on the art of husbandry, designed to promote an acquaintance
with the modern improvements in agriculture, together with remarks on
gardening and a treatise on the management of bees. Hartford, [Conn.],
Samuel G. Goodrich iv,[3],224p   C I
     Bees p. 145-211, 215-24
     1821 Middletown, [Conn.], Clark & Lyman, Print   E
79   BUTTEL-REEPEN, H. v.   1907   Are bees reflex machines?
Experimental contribution to the natural history of the honey-bee.
Transl M.H. Geisler. Medina, A.I. Root Co 48p   C I M X
     1917 2nd ed Natural history of the honeybee; or, Are bees reflex
     machines?   C L M T
80   BYRN, M.L.   1868   The farmer's friend and home companion,
comprising a treatise on the management of bees. Also numerous receipes
of great value to farmers, manufacturers and others. Also---a lecture
on tobacco. N.Y., The Author c1866 18,4,vip   W

81   CALE, G.H.   1924   Every step with package bees. Every Step
Ser. 6 Hamilton, Amer. Bee J. 6p illus
82   CALE, G.H.   1939   Coaching service for the Modified Dadant
Hive. Hamilton, Dadant & Sons 24p illus
    1944 27p
    [See Amer. Bee J. 61(3): 95 (1921)]
83   CALE, G.H., Jr.   nd   Beginning with bees. [5 booklets]
(1) The honey bee colony. 15p; (2) Package bees. 14p; (3) The honey-
flow. 15p; (4) Fall management. 7p; (5) Spring management. 15p
Hamilton, Dadant & Sons
84   CALE, G.H., Jr.   1949   Beekeeping for beginners. Carthage,
Ill., Hancock County J. 114p illus
    1953;   1956;   1959;   1962
85   CALIF. HONEY ADVIS. BOARD   nd   Gems of gold. 35p recipes
86   CARPENTER, C.G.   1844 2nd ed   The bee manager: with directions
for making and managing the Vermont and Perfect bee hives. Geneva, N.Y.,
Ira Merrill 16,2p   W
    Gives credit to J.M. WEEKS for much of content
87   CENTENNIAL BEE-KEEPERS' ASS.   [1877?]   Wintering bees; How
to do it successfully. Chicago, Amer. Bee J. 30p
    Also in COOK 2nd ed   103
88   CHADWICK, P.C.   1921   Big essentials in successful beekeeping.
[Calif.], Press of Selma Irrigator iv,40p   C E I
89   CHAMBERS, M.D.   1921   Honey how and when to use it. San
Antonio, Tex., Amer. Honey Prod. League 21p recipes   C M T
90   CHASE, A.W.   1873   Beekeeping and bee management. In
Dr. Chase's 2nd recipe book.
91   CLARK, E.H.   1918   Constructive beekeeping. Fargo, N. Dak.,
Ulsaker Printing Co 43p illus   C E I L M T W
92   CLARK, J.B.   1930   Virginia and the mason-bee. N.Y., Duttons
55p illus juv   W X
93   CLARKE, W.F.   1886   A bird's-eye view of bee-keeping. In
verse. Beeton, Ont., Can. Bee J. 60p illus adv   C I M T
94   CLARKE, W.F.   1895   Foul brood and sugar honey. 12p   C
    Defense of author's controversial views
95   CLUTE, O.   1878   The blessed bees by John Allen [pseud].
N.Y., G.P. Putnam's Sons 169p   C E T W
    1879   F I L M X;   1881 3rd ed 172p   M X
    Presentation copy to Langstroth at F
    [See Bee World 47(1): 45-46 (1965)]
96   COHN, D.J.   1924   Invertase in honey. Dissertation. N.Y.
38p bibliogr   B F X
97   COLEMAN, G.A.   nd   The amateur beekeeping crank. A pocket
handbook for beginners in beekeeping. 12p typewritten   E
98   COLEMAN, M.L.   1939   Bees in the garden and honey in the
larder. N.Y., Doubleday, Doran, Inc. x,132p illus   C E I L M W X
    Recipes and personal account of experiences keeping bees
99   COLLINGS, PENRY & GREEN   nd   Honey recipes. 76p
100   COLVIN, R. (comp)   1859   A small treatise containing some
important facts with a few practical directions for managing the bee and
hive, compiled by permission from LANGSTROTH on the honey bee. Baltimore,
Forrester & Briingman 48p illus   I
    1860   I
101   COMSTOCK, A.B.   1905   How to keep bees. A handbook for the
use of beginners. N.Y. & London, Doubleday, Page ix,228p illus B E M W
    1906   C;   1907   E I X;   1909   E M;   1911   F;

1914   B I;   1916   I L;   1918;   1919;   1920 rev
ix,230p   B C M T X
102   CONN. BEEKEEPERS ASS.   1952   Connecticut honey and its
uses. Honey recipes and facts about honey. [Rocky Hill] 7p
103   COOK, A.J.   1876   Manual of the apiary. Lansing, Mich.,
W.S. George & Co 59p illus   C I
2,000 of the 3,000 copies printed were sold in one year
1878 2nd ed Chicago, Thomas G. Newman & Son [8],viii,11-286p,adv
C I T X;   1878 3rd ed   I M W;   1879 4th ed viii,11-310p,adv
B C I M W X;   1880 5th ed   B I M W X;   1881 6th ed vi,310p,adv
E M X
1882 7th ed Bee-keeper's guide; or, manual... vi,310p,adv bibliogr
p. 19-23   C E I M T X;   1883 8th ed Columbus, Ohio, Myer's Bro
xiv,347p,adv   C I W X;   1883 9th ed Lansing, Mich. xiv,348p,adv
C M;   1884 10th ed   B T;   1884 11th ed   B I X;   1884 12th ed
C;   1888 13th ed xv,461p,adv glos p. 441-47   B C I X;   1891
14th ed   C E F M X;   1894 15th ed Chicago, George W. York & Co
E I M T X;   1899 16th ed   B C M X;   1902 17th ed xi,543p
B E I M X;   1904 18th ed   X;   1910 19th ed   B E I L M T X
104   COOK, E.H.   1887   G.M. Doolittle's method of rearing queens.
Also a short sketch of the author, and few selections from his writings.
Andover, Conn., The Author 30p illus   M X
105   COOPER, J.F.   1848   The oak openings; or, The bee-hunter.
In two volumes. N.Y., Burgess, Stringer & Co. Vol I 230p; Vol II 228p F
106   CORNELL APIS CLUB   1924-25   Biographies of beekeepers. mimeo
107   COTTON, C.B.   1906   Guide for constructing controllable beehives and fixtures. Gorham, Maine [8]p illus   E
108   COTTON, L.E.   nd   Drawings and specifications for Mrs. L.E.
Cotton's controllable bee hive. 6p   I
An ed 8p   I;   1883 11p   C;   1884 8p   M;   1885
12p   M;
1886 Annual circular. The controllable bee hive and new system of
bee management. 32p   I E
Copy at E has letter tipped in
109   COTTON, L.E.   1880   Bee keeping for profit. A new system of
bee management. 125p   E I T W X
An ed 143p   E;   1883 2nd ed West Gorham, Maine 150p illus
C I M W X;   1891 3rd ed 146p   C;
1906 4th ed Rev by COTTON, C.B. Standish, Maine, H.B. Hartford
42p   E
[See Gleen. Bee Cult. 5(5): 121 (1877); 7(11): 427 (1879); 9(2):
75-76 (1881)]
110   [COTTON, W.C.]   1841   A short and simple letter, from a
conservative bee keeper. Boston, Charles P. Bosson; N.Y., George C.
Thorborn 24p illus   C F I M X
2nd Amer ed   F
Publ England 1837 when author was a student at Oxford. Amer. Mus.
Natural Hist., N.Y. copy inscribed to H.A. King, July 20, 1871: "This
book is a curiosity...only 25 printed...It was the enlargement of a
lecture...and was written in one night...it forms the [introduction]
of my new book."
111   [Cross, J.H.]   [1851]   The hive and its wonders. Written
for the Amer. Sunday-school Union. Philadelphia 126p illus   C I M X
1853 2nd ed London, The Religious Tract Soc.   B M
112   CROWDER, D.E.   1952   The flying nation the story of the
bees. N.Y., Roy Publ 156p illus juv
Publ England

113   CUNNINGHAM, W.F.   1879   Practical bee culture, from personal observation and practical experience. Seymour, Ind., Democrat Steam job rooms 90,2p   C
114   CURTIS, G.DeC.   1948   Bees' ways. Boston, Houghton Mifflin Co xii,240p illus
115   CUTTING, J.A.   1844   A short treatise on the care and management of bees, and the construction of the changeable bee-hive. Newbury, Vt., L.J. McIndoe 12p illus   M
   1847 Boston 16p [Copy at Amer. Antiquarian Soc., Worcester, Mass.]
   1849 Manchester, Vt., The Inventor 16p   X
116   DADANT & SONS   nd   How to remove supers of honey with acid. Hamilton, Amer. Bee J. [3]p illus
117   DADANT, C. & DADANT, C.P.   1881   Extracted honey; harvesting handling, marketing. St. Louis, G.A. Pierrot 24p illus   I M
118   DADANT, C.P.   1914   The bee primer for the prospective beekeeper. Watertown, Wis., G.B. Lewis Co. 21p illus   C E M W
   1921 Milwaukee, Wis. Printing Co   E I;   [1927]   I;
   1937 Hamilton, Amer. Bee J. 30p   B E F M T W X
119   DADANT, C.P. (comp)   1916   Facts about honey [includes recipes]. Hamilton, Amer. Bee J. 16p   B W
   Various repr
120   DADANT, C.P.   1917   First lessons in beekeeping. Hamilton, Amer. Bee J. xii,167p illus   F I T W X
      1918   C;   1919   B I M;   1919 4th ed   C E T X;   1924;
   1926 6th ed   B I T;   1928 7th ed   I;   1930 8th ed   C L;
   1931 9th ed   C I X;   1934 10th ed   C;   1935 10th ed   B;
   1937 12th ed   T;
   1938 Rev by DADANT, M.G. & DADANT, J.C. 127p   B M T W X;   1941;
   1943;   1945;   1946 rev;   1947;   1951 rev;   1952;   1954;
   1957 rev;   1960
   Editions are minor revisions
121   DADANT, C.P.   1920   Dadant system of beekeeping. Hamilton, Amer. Bee J. vi,115p illus   B C E F I L M T X
   1932 2nd ed x,117p   B E I W;   1933 2nd ed   C I;   1934
   10th ed 167p   I
   1922 Le systeme Dadant en apiculture. Quebec, Laflamme 138p illus
                                                             B C L M T
   Spanish, Italian (1923), Russian transl
122   DADANT, C.P. & PELLETT, F.C.   1924   Every step in transferring bees. Every Step Ser 3. Hamilton, Amer. Bee J. 8p illus
123   DADANT, M.G.   1919   Outapiaries and their management. Hamilton, Amer. Bee J. 124p illus   B C E F I L M T W X
   1932 2nd ed 126p   B C E I M W X
124   DANZENBAKER, F.   1902 9th ed   Facts about bees. Management of Danzenbaker's hive for comb honey. Medina, A.I. Root Co 63p illus
"Manufacturer preface"   I
   10th ed   E;   1903 12th ed Rev by ROOT, H.H. & ROOT, E.R. 61p
   1903 13th ed   I M X;   1907 14th ed   ROOT, E.R. The Bee-
   keeper's ten cent Lib 21 60p   C I M W X;   1912   M
125   DARBY, G.   1967   Jerry finds bees. Austin, Tex., Steck-Vaughn 48p illus juv
125a   DAVIDSON   nd   Honey recipe book. 36p
126   DAVIS, H.D.   nd   The beekeepers' pocket companion.
127   DAVIS, S.   [1853?]   S. Davis' platform bee-hive. With instructions. Patented July 26, 1853. Claremont, N.H., Power-press office 11p illus   C

Sylvester Davis' bee feed for "best box honey" on added page with the
admonition to "remember the bonds you are under to keep this a secret".
128   DAVIS, S.B. & GRACE, H.M.   1943   A story for boys and girls
and many new honey recipes for their mothers. N.Y. & Chicago, Mentzer,
Bush & Co. 32p illus
129   DAWSON, A.   nd   Honey bees and their ways. Des Moines,
Iowa, Blackhurst Book Co   I
130   DAY, T.   1795   The history of little Jack, embellished with
pictures; also the history of the little queen, and the natural history
of the silk-worm and bee. Boston, Sold by W. Spotswod juv Bees p.61-78
131   DAYTON, C.W.   1890   The queen-restrictor. Detailing the
only feasible method of brood-nest inversion, contraction, and exclusion---
by the use of brood frames in combination with queen-excluding metals.
Bradford, Iowa, The Author 53p illus   W
     Excluder between frames with two laying queens and one virgin
132   DEANE, S.   [1797 2nd ed]   The New-England farmer. Boston,
Wells & Lilly 532p
     1822 3rd ed Bees p. 25-39   M
133   DEMUTH, G.S.   1923   American foul brood conquered. Medina,
A.I. Root Co 15p   C
     Enthusiastic report on Dr. J.C. Hutzelman's alcohol-formalin method
     for sterilization of combs.
134   DEMUTH, G.S.   1924   Management of out-apiaries. Right Way
Ser. Medina, A.I. Root Co 36p illus   C I T X
135   DEMUTH, G.S.   1925   How to produce comb honey. Medina,
A.I. Root Co 30p illus   C
     1932 .....and swarm control. 29p   M;   Various other ed
136   DEMUTH, G.S.   1941   Five hundred answers to bee questions
pertaining to their behavior and relation to honey production. Medina,
A.I. Root Co 104p
     1942;   1943;   1947;   1948 [2nd ed];
     1955 3rd ed Rev & ed by BARTH, W. 96p
137   DENNIS, L.   [1954]   The story of honey. Toronto, F.R.
Halliday Studio [20p] illus juv
138   DESBOROUGH, J.G.   1879 2nd ed   Lecture on bee-keeping.
Stamford, Conn.
·139   DEYELL, M.J.   1935   How to build up an apiary with combless
package bees. Medina, A.I. Root Co 16p illus   B X
     Various other ed How to .....with package bees.
140   DIAMOND MATCH CO   1915   Cottage bee-keeping. Chico, Calif.
26p illus   I
     nd 31p price changes inserted   C I
141   DILLINGHAM, J.   1853   Bee-keepers' companion, and domestic
economy of the honey bee, dictated by many years of practical operations,
based upon the true principle of the nature, economy and habits of the bee,
peculiar to myself, and with especial reference to my section bee-hive.
Lewiston, [Maine?], William H. Waldron 62p   W
142   DINES, A.M. & DALTON, S.   1968   Honeybees from close up.
N.Y., Thomas Y. Crowell Co 114p illus
143   DIXON, M.   1950 2nd ed   Honey food for life and health. N.Y.,
Mac Richard Publ 47p
     New ed N.Y., Ideal Health Books, Inc.
144   DODDRIDGE, J.   1813   A treatise on the culture of bee.
Ohio 32p
145   DOERING, H.   1962   A bee is born. Transl D.S. Cunningham.
N.Y., Sterling Publ Co 96p illus juv

146   DOOLITTLE, G.M.   1879   Description of the hive I use. Chicago,
Thomas G. Newman & Son 16p illus   E
    An ed .....and my management of bees.   E
    [1882] 15p   E I M
147   DOOLITTLE, G.M.   [1889]   Method of rearing early queens.
Queen Breeders J. Pub 1 Marlboro, Mass., E.L. Pratt 8p   C
148   DOOLITTLE, G.M.   1889   Scientific queen-rearing as practically
applied; being a method by which the best of queen-bees are reared in
perfect accord with nature's ways. Chicago, Thomas G. Newman & Son
163p illus small errata sheet tipped in   C M
    Copy with "appendix" tipped in (numbered 1-6,2p)   T
    Copy with "appendix" bound in   M
    Repr p. 131, line 5: corrected from "customers and neighbors" to
    "neighbors of this customer". Adv for 38,000th of Root's ABC (42,000th
    was next ed) 169p   E I W
    [1891?] repr Adv for 52,000th of Root's ABC   C
    1899 2nd ed "Production and care of comb honey" added on p. 117-126
    126p   B C I M X
    1901 3rd ed   E I M;   1909 4th ed George W. York & Co   B E L T X;
    5th ed   X;   1915 6th ed Hamilton, Amer. Bee J. Paper & hardcover
    B C E I L M T X
149   DOOLITTLE, G.M.   1908   A year's work in an out-apiary or
an average of 114 1/2 pounds of honey per colony in a poor season, and
how it was done. Medina, A.I. Root Co 61p illus   I M T
    1908 2nd ed   C E F;   1909 3rd ed   B I M;   1910 3rd ed   L T;
    1913 4th ed Management of out-apiaries... 72p   B C E I L X;
    1922 5th ed 65p
150   DOUGLASS, B.W.   1921   Every step in beekeeping; a book
for amateur and professional. Indianapolis, [Ind.], Bobbs-Merrill Co
177p illus   B C E F I L M T W X
151   DROWN, W.   1824   Compendium of agriculture or farmers
guide...with the aid and inspection of S. Drown, M.D. Providence, [R.I.],
Field & Maxcy 288p   M
    "Bees" p. 248-52
152   DUAX, I.W. & KIRSCHBAUM, B.H.   1943   Buzzing around with
the honey bees. Chicago, Consolidated Print & Publ Co 64p illus bibliogr
153   DU BOIS, R.   1955   The miracle of royal jelly. Transl
C. Demrick. Hartford, Conn., L.R. Smith & Co 95p illus
154   DU COUEDIC, P.   1829   The pyramidal bee-hive: a plain and
natural method of preserving and perpetuating the population of bees,
and of receiving annually, from each family, a box full of wax and pure
honey, without disturbing the bees, or destroying the couvain; and the
art of restorying hives, (whose population has perished) by hatching the
eggs, remaining in the cells, by the heat of the sun. Also, the art of
converting honey into white, inodorous sugar, and of making hydromel,
sirups, &c. a work useful to farmers. Transl S. Dinsmoor. Philadelphia,
Carey, Lea & Carey 103p illus   M W
    Publ in French 1813
155   DUTCHER, A.P.   1842   A treatise on the natural history
and management of honey-bees, with a description of a new and improved
bee-hive. Pittsburgh, The Author 160p   C
156   DZIERZON, J.   1866   Besste bienenzuchts-methode nach
pfarrer Dzierzon, enhaltend 1. Die aufsätze des vorstehers des Schlesischen
bienenzüchtervereins Wilh. Bruckisch. 2. Das bienenbuch des pfarrer
Dzierzon. 3. Erfahrungen des W. BRUCKISCH durch seine 12 jährige proxis
in Amerika. Hortentown, Tex., W. Bruckisch 216p   X

See W. Bruckish "Bee culture" in Pat Rep for 1860: 268-301 (1861)
for description of Dzierzon hive. Extract in _Amer. Bee J._ 1(12):
277-80 (1861)
157   ECKERT, J.E. & SHAW, F.R.   1960   Beekeeping, successor
to "Beekeeping" by E.F. Phillips [418]. N.Y. & London, Macmillan Co
ix,536p illus
    Append includes important events in Amer beekeeping hist, glos, &
    bibliogr
158   EDDY, H.   1854   Eddy on bee-culture, and the protective
bee-hive; a guide to a successful and profitable method of bee-culture;
the results of many years' experience and observation in bee-keeping.
With an introductory notice by J.V.C. Smith, M.D., mayor of Boston.
Boston, Damrell & Moore x[p. ix missing, x repeated],12-58p [M copy
has 54p] illus   F M W X
    1858 2nd ed xii,15-61p   E M
159   EDGELL, G.H.   1949   The bee hunter. Cambridge, Mass.,
Harvard Univ Press 49p illus
160   EDWARDES, T.   1911 . The lore of honey-bee. N.Y., E.P. Dutton
& Co xix(1),196p   B W
    Publ England
161   EDWARDES, T.   [1923?]   Bee-keeping for all; a manual of
honey-craft. N.Y., E.P. Dutton & Co viii,136p illus   B C F I T
    Publ England, many ed
162   EGGERS, J.F.   1898   Bienenzucht und honig gewinnung nach
der neuesten methode. Grand Island, Nebr., Anzeiger & Herold 51,1p
illus   C I M W
    1900 Lincoln, Nebr., Press Publ Co 91-127p illus   I X
163   ELKON, J.   1955   The honey cookbook. N.Y., Alfred A. Knopf
xv,162,xii p
163a   ELLIS, G.F.   nd   Lessons on the honey-bee for high school
students. Ann Arbor, Mich., Edwards Bro 35p illus mimeo   C
164   ENGLISH, C.H.   nd   Honey, the health sweet. Toronto,
Ont. Honey Prod Co-operative Ltd 12p
165   EVERETT, D.   1905   A clergyman and his bees by Clericus
[pseud]. Medina, A.I. Root Co 19,1p illus   C
    An ed nd   C F
166   EVRARD, E.   1923   The mystery of the hive. Transl B. Miall.
N.Y., Dodd, Mead & Co iv,369p   B C F I M T W X
167   EWALD, C.   1908   The queen bee; and other nature stories.
Transl G.C. Moore Smith. N.Y. 125p illus juv   B C M
168   FALCONER, W.T., CO.   nd   Simplified bee-keeping. Falconer,
N.Y. 36p illus   C E F I M
    See 242
169   FILLEUL, P.V.M.   1851   The cottage bee keeper; or suggestions
for the practical management of amateur, cottage and farm apiaries, on
scientific principles: with an appendix of notes, chiefly illustrative:
By a country curate [pseud]. N.Y., C.M. Saxton 12,119p illus   F I M W
    1856   C I L
    Publ London 1851 The English beekeeper;... many ed   B
170   FISCHER, M.   1926   Honey way menus. Madison, Wis., Honey
Tea Room 61p illus   B C M T
171   FLANDERS, W.A.   1854   The honey-bee and hive. Northfield,
Vt., Woodworth & Gould 16p illus   C
    1861 4th ed The honey bee, and movable comb hive; showing the new
    system of bee-keeping. Cleveland, Ohio, B. Franklin 24p   W
    Pat March 6, 1860

172   FLANDERS, W.A.   1863   Sweet home; or, nature's bee-book.
Vol 1, No 1, semi-annual. Cleveland, Ohio, Nonpareil Steam Print. House,
Viets & Savage 1,63,1p illus   W
    1865 2nd new ed 64p   I
    1867 3rd new ed Nature's bee book, a practical treatise, calculated
    to assist the bee-keeper in overcoming the difficulties and mysteries
    of bee-keeping and insure profitable returns for labor and capital
    invested. Mansfield, Ohio, L.D. Myers & Bro   W
173   FLICK, H.H.   1873   The busy bee, an illustrated manual on
scientific and practical bee culture. Somerset, Pa., Herald Print Co
15p illus   W X
    Included price list of equipment offered for sale
174   FLOWER, A.B.   1925   Beekeeping up to date. N.Y., Toronto
etc., Cassell & Co 110p illus   L T
    Publ England many ed   B
175   FORTUNE, M.S.   nd   Beginners' bee briefs. Mayhew, Miss.,
Stover Apiaries 21p illus   C
176   FOSTER, O.   1886   How to raise comb honey. Describing
improvements in methods resulting from ten years practical work, and
extensive experiment. Mt. Vernon, Iowa, Edson Fish 14,2p illus   E I W
    Offered price-list of supplies
177   FOSTER, V.E.   1960   Close-up of a honeybee. N.Y., Young
Scott Books 64p illus juv
178   FRANCE, N.E.   1901   Diseases of bees and legislation.
Platteville, Wis., Journal job rooms 28p illus   M
179   [FRANCE, N.E. (comp)]   [1903?]   Bees and horticulture;
their relations mutual. Platteville, Wis., Nat Bee-keepers' Ass 12p C E
    Includes 383 & 526
180   FRANCE, N.E. (comp)   1904   Legal rights 1904. Platteville,
Wis., Nat Bee-keepers' Ass 37p   M
    Includes 369
    1910 Bee-keepers' legal rights. 59p   M T X
    Includes 179
181   FRANK, A.   1848   Manual on the management of bees: giving
a description of the bee, its disposition and habits, and designed to
describe and accompany the self-protecting bee hive; illustrated.
Busti, N.Y., [Jamestown J] 72p illus   C T W
182   FREY, N.A.   1932   Apis the hive bee. N.Y., Frederick A.
Stokes Co 140p illus juv   C I W
183   FRISCH, K. von   1950   Bees: their vision, chemical senses,
and language. Ithaca, N.Y., Cornell Univ Press xiii,119p illus bibliogr
    Many repr. Paperbound, Great Seal Book;   1971 rev 176p
184   FRISCH, K. von   1955   The dancing bees; an account of the
life and senses of the honey bee. Transl D. Ilse of 5th rev German ed
(1953). N.Y., Harcourt, Brace 183p illus
    1966 2nd ed Transl of 7th German ed (1964)
185   FRISCH, K. von   1965   Tanzsprache und Orientierung der
Bienen. N.Y. & Berlin, Springer-Verlag vii,578p illus bibliogr
    1967 The dance language and orientation of bees. Transl L.E. Chadwick.
    Cambridge, Mass., Harvard Univ Press xiv,566p
186   FRISCH, K. von   1967   A biologist remembers. Transl
L. Gombrich. N.Y., Pergamon Press 202p illus
187   FRY, R.K.   1938   Bumblebuzz. N.Y., E.P. Dutton & Co 25p
illus juv   B W
188   GILMAN, A.   1929   Practical bee-breeding. N.Y. & London,
G.P. Putnam's Sons xvi,264p illus   B C E F I L M T W X
    Publ England 1928 with index

189   GILMORE, J.M.   1854   Gilmore's patent bee-house and bee-
hive. N.Y., C.M. Saxton 16p illus   W
    1856 Windsor, Vt., Vermont Chronicle Press 23p illus   I M
190   GIVEN, D.S.   [1883]   Directions for the use of Given's
foundation press and wiring machine. Hoopeston, Ill., D.S. Given & Co
12p   I
191   GOUDEY, A.E.   1960   Here come the bees! N.Y., Scribner
94p illus juv
192   GRACE, H.M.   1947   Honey, the oldest sweet. Chicago,
Phoenix Metal Co recipes
193   GRACE, H.M.   1956   More favorite honey recipes. Madison,
Wis., Amer Honey Inst 64p illus
194   GRAFFAM, G.S.   1954   The fragrant city of wax. Augusta,
Maine, Kennebec J 112p illus
195   GRANER, E.W. nd Honey production manual. N.Y., Long Island
Beekeeper's Ass 17p illus mimeo   C
196   GREEN, S.N.   1907   .Bee-breeding, a little monograph on a
neglected subject. Medina, A.I. Root Co 42p   C I M T W X
197   GRINNELL, E.   1903   A morning with the bees. Medina,
A.I. Root Co 14p illus
198   GROUT, R.A. (ed)   1946   The hive and the honeybee; a new
book on beekeeping to succeed the book "Langstroth on the hive and the
honeybee" [284]. Hamilton, Dadant & Sons 16,633,18(index)p illus
    1949 rev xviii,652p;   1954;   1960;   1961;
    1963 rev xvii,556p
    1969 Russian transl;   1971 German transl
199   GROUT, R.A.   [1950]   Pollination - an agricultural practice.
Hamilton, Dadant & Sons 23p illus
    [1951] repr Planned pollination - an agricultural practice. 24p
    1953 23p
200   GROVES, W.   1837   Recommendations and certificates of
Groves' patent fortified, transparent, Royal Bee Palace, ... nature and
habits of these wonderful working insects. Pittsburgh 16p [Copy at
Amer. Antiquarian Soc., Worcester, Mass.]
201   HALL, W.M.   1840   The apiarian, or, a practical treatise on
.the management of bees; with the best method of preventing the depredations
of the bee moth. New Haven, [Conn.], Hitchcock & Stafford 48p   C M W
    1841   C F I
202   HALLOCK, T.P.   nd   Back-yard bee-keeping. Medina, A.I. Root
Co 18p illus
203   HAMLIN, T.B.   1871   A practical treatise on improved bee
culture, containing plain and concise directions for artificial swarming,
hiving, transferring bees from the common to the movable frame hive.
Feeding, wintering and care of bees at all times of the year, how to obtain
the largest amount of surplus honey, how to protect against the bee moth,
etc., etc., with full directions for the use of the movable-comb bee hive.
Nashville, Tenn., Warfedale Press of Paul & Tavel 76p   C
204   HAND, J.E.   1911   Beekeeping by twentieth century methods.
Medina, A.I. Root 59p illus   C E I M T X
205   HARBISON, J.S.   1859   Harbison's California adjustable comb
hive, pat. by J.S. Harbison Jan. 4, 1859. New Castle, Pa., J.M. Kuester
11p   I
    Hive allowed moth to drop out
206   HARBISON, J.S.   1860   An improved system of propagating the
honey bee. Sacramento, Calif., Democratic Standard Office 20p illus M W
    Right to use hive included

207   HARBISON, J.S.   1861   The bee-keeper's directory, or the
theory and practice of bee culture, in all its departments, the result
of eighteen years personal study of their habits and instincts. With an
introductory essay by D.C. Wheeler. San Francisco, H.H. Bancroft & Co
440p illus [J.S.H. mss see Amer. Bee J. 109:352(1969)]   B C E I L M W
208   HARBISON, W.C.   1860   Bees and bee-keeping: a plain,
practical work; resulting from years of experience and close observation
in extensive apiaries, both in Pennsylvania and California. With directions
how to make bee-keeping a desirable and lucrative business, and for shipping
bees to California. N.Y., C.M. Saxton, Barker & Co 288p illus
                                                            B C E I L M T
209   HARRIS, L.D. & HARRIS, N.D.   1956   Hummer and buzz. Boston,
Little, Brown 55p illus juv
210   HARRISON, C.   1903   The book of the honey bee. London &
N.Y., J. Lane vii,132p illus   B L M T W
211   HART, A.H.   1876   Fifty years of beekeeping. Appleton,
Wis.   [From Richter list]
212   HASLUCK, P.N. (ed)   1905   Beehives and bee keepers' appliances.
London, N.Y., Toronto etc., Cassell & Co 160p illus   T
    Publ England, numerous repr
213   HAWES, J.   1964   Bees and beehives. N.Y., T.Y. Crowell,
Collier & Macmillan illus juv
214   HAWES, J.   1964   Watch honeybees with me. N.Y., T.Y. Crowell,
Collier & Macmillan illus juv
215   HAWKINS, K.   1920   Beekeeping in the south. A handbook on
seasons, methods and honey flora of the fifteen southern states. Hamilton,
Amer. Bee J. 120p illus   B C F I M X
216   HAWKINS, K.   1934   A hobby that pays. Watertown, Wis.,
G.B. Lewis Co 16p illus
217   HAWKINS, K.   1935   The beekeepers guide. 79p illus   M
    Author's notation: "Publ for Sears Roebuck Co no publicity please."
    1936 3rd ed   I X;   1938 4th ed   X;   1940 7th ed: 1941 8th e
    1942 9th ed;   1942 10th ed;   1943 11th ed;   1943 12th ed;
    1944 13th ed;   1946 14th ed
218   HAWKS, E.   1912   Bees. N.Y., Thomas Nelson & Sons 120p
illus juv
    Publ England. Bees shown to children.   B
219   HEAVILIN, J.   1964   Bees and wasps. N.Y., Macmillan illus
juv
220   HEDDON, J.   [1885]   Success in bee-culture, as practiced
and advised by James Heddon. Dowagiac, Mich., Ed & Publ Dowagiac Times
128p illus   C E I L M T W
221   HEDDON, J.   1890   Heddons net price list of useful implements
for beekeepers. The new hive and system. Heddons catalogue for 1890.
Dowagiac, Mich., The Author 40p illus   M
222   HEDDON, J.   1895   A new system of management in beekeeping;
and new hive specially adapted thereto. Dowagiac, Mich. [21p]   M C
223   HEWITT, P.J., Jr. (comp)   1956   The Hewitt collection of
apiculture, Litchfield Hist. Mus. Conn. Beekeepers Ass Inc 37p mimeo
224   HICKS, J.M.   1875   The North American bee-keepers' guide.
Lafayette, [Ind.], J.L. Cox & Bro viii,106p illus   W
    Sold hives, bees and Italian queens
225   HILBERT, J.E.   1932   Secrets of cut comb honey. Traverse
City, Mich. [18p] illus
    Sold equipment and supplies for cut comb.

226   HILTON, G.E.   [1887?]   How I produce comb honey. Fremont,
Mich., Fremont Indicator Job Print 12p illus   E M X
227   HLEBOWICZ, Z.   1920   Pszczoly i pszozelnictwo. Stevens
Point, Wis., Bracia Worzellowie 405p illus   L M W
228   HOBART, J.   1866   Treatise on bees and bee culture, comprising
a history of the bee, statement of a new theory in respect to the formation
of comb, criticism of ventilated hives, description of the pirate bee-hive,
and directions for the management of the apiary. Monroe, Wis., High &
Booth, Sentinel Office 28p illus   M
229   HOFFMAN, F.A.   1886   Amerikanische bienenzucht. Ein hand-
buch für angehende bienenwirthe. Unter benutzung der besten amerikanischen
und deutschen schriftsteller bearb. von Hans Buschbauer [pseud]. Milwaukee,
Wis., Geo. Brumder 138p illus   C E I M W
230   HOFFMAN, M.   1853   Treatise on bees, their keeping and
management, with directions for hiving, wintering, etc. Portage City,
John A. Brown 8p   W
231   HOOPER, F.A.   1902   Bee-keeping in Jamaica. Medina, A.I.
Root Co 61p illus   .   C E M X
    Tropical beekeeping
232   HOOPER, N.R.   1952   Beehives and apron strings. N.Y.,
Comet Press 114p illus
233   HOOTMAN, H.D. & CALE, G.H.   nd   Busy bees bring bending
branches. Hamilton, Amer. Bee J. 23p illus   C M T X
234   HOWARD, W.R.   nd   New York bee-disease, or black brood.
235   HOWARD, W.R.   1894   Foul brood; its natural history and
rational treatment. With a review of the work of others. Chicago, George
W. York Co 48p illus   B C E I L M W
236   HOYT, M.   1965   The world of bees. N.Y., Coward McCann
254p illus
237   HUBBARD, G.K.   1886   First principles in progressive bee
culture. A book of instructions for handling bees when dividing, transferring, uniting, feeding, wintering, introducing queens, hiving swarms,
destroying fertile workers and moths, stopping robbing, etc., etc. Medina,
Gleanings Print 55p p. 31-55 price list of Hubbard bee-hive illus E M W
    1890 rev 104p p. 71-104 price list   C I;   1892 3rd ed Fort
    Wayne, Ind., The Author 101p p. 71-101 price list   C
238   HUBER, F.   1926   New observations upon bees. Transl DADANT,
C.P. 2v in 1 Hamilton, Amer. Bee J. 230p illus   B C E M W X
    Publ in French 1792 Geneva, other ed
    English transl 1806, 1821, Edinburgh
    1971 repr of 1821 Edinburgh ed New observations on the natural history
    of the bees. Louisville, Ky., Black Stone Press
239   HUNT, G.V.   1963   Four little bees. Salt Lake City, Utah,
Deseret Book Co 47p illus juv
240   HUTCHINSON, W.Z.   1887   The production of comb honey, as
practiced and advised by W.Z. Hutchinson. Flint, Mich., Glove Print
House 45p   C I L M T
241   HUTCHINSON, W.Z.   1891   Advanced bee-culture its methods
and management. Flint, Mich., The Review Print 88p illus   C E I M X
    1902 2nd ed 90p   I M X;   1905 3rd ed 204p,adv   B C E I M T X;
    1911 4th ed 205p   B C E I M T X;   1918 5th ed Medina, A.I. Root
    Co   B C E L T
    Ed & publ of Bee-Keeper's Review started 1888
242   HUTCHINSON, W.Z.   1897   Successful bee keeping. Jamestown,
N.Y., W.T. Falconer Mfg Co 16p See 168   C
243   INTERNATIONAL CONGRESS of BEEKEEPERS 1924   VIIth International
Congress of Beekeepers. Quebec, Charrier & Dugal 500p illus   B M X
    In French & English

-22-

244   JACQUES, D.H.   1858   Domestic animals. Rural handbook No 3.
N.Y., Fowler & Wells 168p beekeeping p. 148-159 illus   W X
   1866 rev WOODWARD, G.E. & F.W.   W
245   JAGER, F.   1929   Bee library of Francis Jager. St. Bonifacius,
Minn. 62,17,5p typewritten   M
   Library sold to Univ of Minn., 1929
246   JARVIS, D.C.   1958   Folk medicine. A Vermont doctor's guide
to good health. N.Y., Henry Holt & Co 182p
   1959 repr
   Transl into 12 languages
247   JARVIS, D.C.   1960   Arthritis and folk medicine. N.Y.,
Holt, Rinehart & Winston 179p
248   JENKINS, J.M.   1898 13th ed   Bee-keeping in Dixie. 61p
illus   I X
   1901 16th ed   L X
   All but 8p concerns supplies
249   JENSEN, A.   1965   Droniest bee in the hive. St. Louis,
Bethany Press juv
250   JENSEN, M.F. (comp)   nd   Honey and its uses in the home.
Pomona, Calif., Southern Counties Gas Co 32p recipes
251   JEPSON, H.H.   [1905]   Outfits for beginners. Boston 25p
   A.I. Root agent
252   JOHNSON, H.M.   1867   Nature's bee book.
253   JOHNSON, H.M.   1872   The farmers' guide to beekeeping.
Being a practical treatise in bee culture and bee management. Ann Arbor,
[Mich.], Courier Steam Print House 223,ivp sheet of errata inserted
illus   W
254   JONES, D.A.   nd   Foul brood, its management and cure.
Beeton, Ont., Can. Bee J. 32p   I
   First major commercial honey producer in Canada. Searched Old World
   for species of bees (in partnership with F. Benton) and bred queens on
   isolated islands of Georgian Bay.
   [See Glean. Bee Cult. 15(23): 892-94 (1887)]
255   JONES, H.   1909   A radical cure for the swarming habit of
bees. Preston, Minn. 21p   C I M W
   1910 25p   L M W X
256   JONES, J.   1843   Manual, for the use of Jones' multiplying
and equalizing hive; containing also hints useful for the management of
bees in all sorts of hives. Ballston Spa, N.Y., J. Comstock 13p,errata C
257   KARABUT, N.   1938   Canadian beekeeping.
258   KAUFMAN, M.   1966   No holidays for honeybees. N.Y., Coward-
McCann 63p illus juv
259   KELLEY, W.T.   nd   How to grow queens for 15¢ each. Paducah,
Ky., The Author 19p illus
   [1941] 2nd ed How to grow queens.
260   KELLEY, W.T.   1955   How to keep bees and sell honey.
Clarkson, Ky., The Author 106p illus
   1958 2nd ed 112p;   1968 5th ed
261   KELLOGG, V.L.   1920   Nuova; or, the new bee, a story for
children of five to fifty...with songs by Charlotte Kellogg. Boston &
N.Y., Houghton Mifflin Co 150p illus juv   C F I M W
262   KELSEY, F.   1835   A practical treatise on the description,
cultivation and management of honey bees. N.Y., H.R. Piercy iv,24p   C F
   1837 Print at 5 Eldrige St.   F
263   KELSEY, W.R.   1847   The apiarian's guide, being a practical
treatise on the culture and management of bees. Syracuse, N.Y., Kinney,
Marsh & Barnes c1846 46p illus   W

```
       1849 4th ed  Philadelphia, William F. Geddes  32p              I
       1866 10th ed  KELSEY, A.  Cincinnati, C.N. Morris  30p          C
264     KENLY, J.C.      1935     Cities of wax.  N.Y. & London, D. Appleton-
Century Co  xv,250p  illus                                      C I L M W X
265     KIDDER, K.P.     1858     Kidder's guide to apiarian science,
being a practical treatise, in every department of bee culture and bee
management.  Embracing the natural history of the bee, from the earliest
period of the world, down to the present time;  giving the anatomy and
physiology of the different species of bees that constitute a colony, etc.
Burlington, Vt., C.B. Nichols; Chicago, Rufus Blanchard  175p  illus
                                                                B C E F M T W

      [1859]  Adv circ for beeware by author inserted                  E
      1858  As 1858 above with ticket pasted over publ:  Boston, Taggard &
      Chase                                                            M
      1865 2nd ed  Secrets of bee-keeping.  Being...  Burlington, Vt., Free
      Press Steam Print Est  151p                                    C I
      1868 4th ed  The Author  192p                              C E F L M X
      1879 with JENKINS  Beekeeping;  being...  New Liberty, Ky., Kidder &
      Jenkins  223p                                                    M
266     KIDDER, K.P.     1863 8th ed     He that wishes to thrive and make
money with ease, should buy Kidder's Pat. Hives, and commence keeping
bees.  Send 50 cents for Kidder's bee book.  Guide to apiarian science.
Which I will forward on receipt of price.  Burlington, Vt.  24p  illus   C
267     KIDDER, K.P.     1863     Secrets of bee-keeping.  Vol 1st
[Burlington, Vt.?], The Author  64p                                    F
268     KIDDER, K.P.    [1865]    Italian bee-book.  [Listed in catalogue
of L.L. Langstroth's bee-library]
269     KILLION, C.E.    1951     Honey in the comb.  Paris, Ill.,
Killion & Sons Apiaries  114p  illus
270     KIMBALL, J.N.    1917     On the trail of the busy bee.  N.Y.,
[Rider Press]  13p  juv                                               W
271     KIMM, S.C.       1922     A tribute to Philip H. Elwood 1847-1922.
Little Falls, N.Y., Little Falls Times, May 18, 1922  4p               C
272     KING, H.A.       1903     The bride of his palace.  Boston, H. Woodman
Press  293p  illus                                                     I
273     KING, N.H. & H.A.      1864     The bee keeper's text book, or
facts in beekeeping, with alphabetical index, being a complete reference
book, on all practical subjects connected with the culture of the honey
bee, for both common and movable comb hives, giving minute directions for
the management of bees, in every month of the year and illustrating the
nucleus system of swarming.  For the Amer. Bee-hive Co of Nevada, Ohio.
Cleveland, Ohio, Vieta & Savage  128p  illus                         C I T
      1866 2nd ed  Buffalo, Franklin Print House  140p                 C
      1867 3rd ed  Printed on cover:  4th ed  Thomas, Howard, & Johnson  E
      1867 5th ed    C M;    1868 5th ed wrappers    C;    1868 6th ed    I;
      1868 8th ed    I;    1869 9th ed  N.Y., H.A. King & Co    C M X;
      1870 10th ed    E;    13th ed  140,2p    I M;    [1871?] 14th ed
      140,12p    I;    1872 21st ed  140,4p  price list    C I M;
      1873 22nd ed  140p,18p  price list wrappers    M;    1873 22nd ed
      140,16p  hard cover    M;    1876 23rd ed  King & Slocum  139,1p B C M;
      1878 23rd ed  139,2p    E I X
      1878 24th ed  rev KING, A.J.  The new bee-keepers' text book.  A.J. King
      & Co  229p                                                C E I L M W
      1879 24th ed  O. Judd Co    B E M X;    1881 25th ed  A.J. King & Co
      C T;    1883 26th ed  King & Aspinwall  228p    C;    1887 26th ed
      Barrytown, N.Y., Aspinwall & Treadwell  228p    I;    1888 27th ed
      John Aspinwall    C E I M
```

1866   Des Bienen-zuchters leitfaden. Cleveland, Ohio, W.J. Schneider
     159p illus   I
274   KIRBY, W. & SPENCE, W.   1846 (from 6th London ed)   An introduction to entomology... Philadelphia, Lea & Blanchard 600p illus
     Publ England 1815
275   KOENIG, M.   nd Bee. Animals at home series. N.Y., Grosset
& Dunlap illus juv
276   KOHNKE, A.R.   1882   Foul brood: its origin, development
and cure. Youngstown, Ohio, Youngstown Publ Co 13p   B C E L M W X
277   KRETCHMER, E.   [1869?]   Winke fur Bienen Zuchter and transl
Intimation to beekeepers.
278   KRETCHMER, E.   [1870]   The beekeeper's guide. 244p
     1872 The American bee-keepers' guide; a new and re-written edition
     of the Bee-Keepers' guide book, a complete manual and reference book
     on all subjects connected with successful bee-culture, in both common
     and movable-comb hives, giving plain directions for the management of
     bees, in every month of the year with alphabetical index, and thoroughly
     illustrating the new system of bee culture with the honey extractor.
     Containing, also, practical directions how to rear Italian queen bees.
     c1862 Chicago, Wabash Steam Print House 244p   B C I M X
279   KROGH, A.   1948   The language of the bees. San Francisco,
W.H. Freeman Co
280   KRUSCHKE BROS.   1874   Rape culture its value as a farm crop
and honey plant, briefly set forth. Berlin, Wis., Berlin Courant Print
23p   I X
281   LAIDLAW, H.H. & ECKERT, J.E.   1950   Queen rearing. Hamilton,
Dadant & Sons, Inc 147p illus
     1962 2nd ed Berkeley, Calif., Univ Calif. Press viii,165p
282   LAMBURN, J.B.C.   1958   A hive of bees by John Crompton [pseud].
Garden City, N.Y., Doubleday x,180p illus
     Publ Edinburgh [1947] The hive.
283   LANCE, P.C.   1938   Queendom of the honey bees. Harrisburg,
Pa., Stackpole Sons viii,110p illus bibliogr juv   C E I L M W X
284   LANGSTROTH, L.L.   1853   Langstroth on the hive and the
honey-bee, a bee keeper's manual. Northampton, Mass., Hopkins, Bridgman
& Co xvi,384p front   B C I L M T X
     1914 repr 1853 ed Medina, A.I. Root Co xvi,13-378p 18 plates from
     2nd ed text on p. 344-78 from p. 458-92 of 2nd ed   B C E I M T X
     1857 2nd ed N.Y., C.M. Saxton & Co 468p append p. 469-79 plates
     and explanation p. 481-510 index p. 511-34   B C F I L M T W X
     1859 3rd ed N.Y., A.O. Moore & Co 408p 23 plates adv for Langstroth
     hive p. 407-08   C E I L M W
     R. Colvin's ed as above   C W X;   1860 xii,386p   I;   1860
     N.Y., C.M. Saxton, Barker & Co; San Francisco, H.H. Bancroft & Co   C;
     1860 C.M. Saxton, Barker & Co 388[6]p index p. 387-405 adv p. 407-08
     C E;   1860 409p adv p. 411-12   F L T W;   1860 C.M. Saxton,
     Barker & Co; H.H. Bancroft & Co   C
     [See Bee World 48(3): 110-12 (1967)]
     1861 C.M. Saxton, Barker & Co   I;   1862 C.M. Saxton   E I M;
     1863   F I M W;   1865 N.Y., Orange Judd   I M W X;   1865
     Philadelphia, J.B. Lippincott   E M T W;   1866   I W;   1868
     E F I M T;   1870   E C M X;   1871   B I M;   1872   I;
     1875 4th ed xii,409p   C I
     1876 [From J.E. Hawkins catalog];   1878   I;   1879   E I;
     1881   M T;   1883   C I M;   1884   I M
     1889 rev by Chas. Dadant & Son Hamilton c1888 xv,521,v,xiiip
                                                         C F I L M T W X

```
1890    B M;        1892    C;
1893 2nd ed of rev ed     I X;      1896 3rd ed      I M T;
1899 4th ed     I T X;      1900 5th ed      I M;      1902 6th     C I T X;
1904 7th ed  Photo & biography of C.Dadant inserted after v1      C M;
1905 8th ed     F I;      1907 rev ed  x,575p     C I M X;      1908     X;
1909     B;      1911    E T;      1913    C E I T;      1914     B T;
1915     B I T;      1918    I T;      1919    C E F L;
1922 21st ed  Rev by DADANT, C.P.  ix,438p               C E I L M T W X
1923 22nd ed     B C F I X;      1927 23rd ed     B C E I W X
[See Bee World 48(4): 133-43 (1967)]
Transl French, Geneva 1891, 1908;  Russian 1892;  Italian, Torino 1928;
Polish, Lwow 1930;  Spanish, Barcelona 1943
```
284a   LANGSTROTH, L.L.   1888   Handling bees. Hamilton, Chas.
Dadant & Son 24p   C E I M
      p. 185-205, 519-21 of Dadant's rev of Langstroth's "Hive and the honeybee"
285   [LANGSTROTH, L.L.]   [1895?]   Catalogue. Bee-library of the
late Rev. L.L. Langstroth. With Masonic works from library of Robert H.
Thomas. 17p L.L.L. library p. 1-6   F
      [List of titles published in Amer. Bee J. 35(48): 765 (1895)]
      [See also F. Naile p. 193-208; Bee World 53(1): 22-27 (1972)]
286   LATHAM, A.   1908   The Latham let-alone-hive. Norwich, Conn.
2p 5 fig mimeo   X
287   LATHAM, A.   1946   Bee management helps for new beekeepers.
Conn. Beekeepers Ass 7p
288   LATHAM, A.   1949   Allen Latham's bee book. Hapeville, Ga.,
Hale Publ Co 200p illus glos
289   LATROP, H.   1927   The yankee abroad. Bridgeport, Wis.,
Record Print 93p   E I M
      Beekeeping in British Isles p. 70-76
290   LAVINE, S.A.   1958   Wonders of the hive. N.Y., Dodd, Mead
Wonder Books 92p illus juv
      1962
291   LESTOURGEON, E.G.   1924   Every step in bottling honey. Every
Step Ser. 6 Hamilton, Amer. Bee J. 15p illus
292   LEWELLEN, J.B.   1953   The true book of honeybees. Chicago,
Childrens Press 45p illus juv
293   LEWIS, [G.B.], CO.   nd   Bee pranks. Chicago, Whites Class
Adv Co 16p
      2nd ed Watertown, Wis., Glaus & Jaeger
      Newspaper clippings concerning bees
294   LEWIS, G.B., CO.   nd   Points for beginnings. Watertown,
Wis. 12p
295   LEWIS, G.B., CO.   1919-22   How to booklets.   B C E M X
    1) How to manage bees in spring:... 8p
    2) How to control swarming:... 8p
    3) How to buy queen bees:... 8p
    4) How to start bee keeping:... 8p
    5) How to locate an apiary:... 8p
    6) How to feed bees:... 8p
    7) How to transfer bees:... 8p
    8) How to care for nuclei:... 7p
    9) How to unite bees:... 8p
   10) How to winter bees outdoors:... 7p
   11) How to use comb foundation:... 8p
   12) How to care for package bees:... 8p
   13) How to extract wax from combs:... 8p

14) How to use an observation hive:... 8p
15) How to organize beekeepers.
16) How to control wax worms:... 8p
17) How to make increase:... 8p
18) How to produce section comb honey:... 8p
19) How to manage bees in fall:...
20) How to remove bees from trees:... 7p
21) How to produce extracted honey:... 8p

296   LEWIS, G.B., CO.   1921   Profit in beekeeping. Watertown,
Wis. 23p illus   B I M X
    1923
297   [LEWIS, G.B., CO.]   1927   How to produce honey. Watertown,
Wis. 32p illus   M
298   LEWIS, G.B., CO.   1928   Honey recipes. 22p illus   M
299   LEWIS, G.C. (comp)   1924   How to market honey. Madison,
Wis., Democrat Print Co; Watertown, Wis., G.B. Lewis Co 84p illus
                                                          C F I M W
    1926 2nd ed   40p   X;   1927 3rd ed   E;   1929 4th ed   M X;
    1931 5th ed   C X
Collection of brief articles "told by 40 leading beekeepers"
300   LINDAUER, M.   1961   Communication among social bees.
Cambridge, Mass., Harvard Univ Press; London, Oxford Univ Press vii,143p
illus bibliogr
    Harvard Books in Biology No. 2
    1967 N.Y., Atheneum Press [paper back]
301   LLOYD, J.W.   1911   A treatise on apis (the bee), tela aranaae
(cobweb), spongia and cantharis. Cincinnati, Lloyd Bro 48p illus   C M
    Drug treatise No. 21, bees p. 3-20
302   LOCKARD, J.R.   1908   Bee hunting. A book of valuable
information for bee hunters - tells how to line bees to trees, etc. St.
Louis, A.R. Harding Publ Co 72p   C E M T W X
    1908 Columbus, Ohio   F I L M T;   1932   T;   1936   B C
303   LO PINTO, M.   1957   Eat honey and live longer. N.Y., Twayne
Publ, Inc 174p glos
304   LOS ANGELES CITY SCHOOL DISTRICT   1934   Materials and
activities relating to honey bees. Natural science unit, grades 3-4.
Los Angeles, Calif., City School District, Div. of Curriculum & Instruction,
Elementary Curriculum Sect. 27p   W
305   LOVE, J.W.   nd   The business of beekeeping. Medina, A.I. Root
Co 31p
306   LOVE, J.W.   1915   The airline honey book. Medina, A.I. Root
Co 64p illus   C M W
    1921   M;   1922   48p   T
307   LOVELL, H.B.   1956   Honey plants manual. Medina, A.I. Root
Co 64p illus
308   LOVELL, J.H.   1918   The flower and the bee. Plant life and
pollination. N.Y., Charles Scribner's Sons xvii,286p illus C F I L M T W
    1919 London, Constable   B
309   LOVELL, J.H.   1926   Honey plants of North America. Medina,
A.I. Root Co 111,408p illus   B C E F I L M T W
310   LUNT, H.   1899   As the wild bee hums. Cincinnati, Editor
Publ Co 11,178p illus
311   LUTTINGER, P.   1922   Bees' honey in substitute infant
feeding. N.Y. 10p
312   LYON, D.E.   1904   Pleasure and profit of honey production.
Medina, A.I. Root Co 17p illus   M

313   LYON, D.E.   1909   Biggle bee book. A swarm of facts on
practical beekeeping, carefully hived, by Jacob Biggle ⌈pseud⌉. Philadelphia,
Wilmer Atkinson Co 136p illus   E M W X
    1913 2nd ed   C F I M T
314   LYON, D.E.   1910   How to keep bees for profit. N.Y.,
Macmillan Co xii,329p illus   B F I L M W X
    1910 (June)   E C;   1912   M;   1913   T X;   1914;
    1915   M;   1916;   1917   B;   1918   W;   1920;   1927
315   MC CALEB, W.F.   1917   Happy; the life of a bee. N.Y. &
London, Harper & Bro 120p illus juv   C F I M W
316   MC CORMICK & CO   nd   "Flavor-it" the Bee Brand way.
Baltimore 31p recipes
317   MC CORMICK, M.   1960   The golden pollen. Yakima, Wash.,
Yakima Bindery & Print Co 1v,160p illus
318   [MC DOUGALL, G.P.]   1879   Secrets of bee-keeping, or how to
make money with the honey bee. Chicago, F. Andrews 68p illus
    1888 63p
    [See Amer. Bee J. 15(8): 341 (1879)]
    Adv for vegetable cathartic, German antiphlogistic plaster and new
    Improved Hive sold by Andrews on back cover
319   MC INTYRE, F.   1938   Children of the golden queen. N.Y.,
E.P. Dutton & Co 80p illus juv   C I L W X
    An ed Eau Claire, Wis., Cadmus Books, E.M. Hale & Co
320   MACE, H.   ⌈1921⌉   A book about the bee. N.Y., E.P. Dutton
& Co 138p illus   B C L T
    Publ England [1921], other ed
321   MAETERLINCK, M.   1901   The life of the bee. Transl A. Sutro.
N.Y., Dodd, Mead & Co 427p bibliogr   E F I M W
    1904   X;   1905   F;   1912 265p   F;   1913 427p   I;
    1915   W;   1916   X;   1924   C;   1926   X;   1927;
    1928   L;   1929;   1936   W;   1942
    1954 New Amer Libr Mentor Book 111 168p
    1964
    Publ England, many ed   B
322   MAETERLINCK, M.   1906   The swarm from the "Life of the bee".
Transl A. Sutro. N.Y., Dodd, Mead & Co 113p illus   C T W
323   MAETERLINCK, M.   1919   The children's life of the bee...
selected and arranged by A. Sutro & H. Williams. N.Y., Dodd, Mead & Co
192p illus   I L W
    1920   T;   1940   X
    Publ England   B
324   MAGUIRE, V.C.   [1947]   Honey Sunday 'n' Monday 'n' all
through the week. Guelph, Ont., Ont. Beekeepers Ass 24p   C T W
    Recipes also publ by Ont. Honey Prod Co-op entitled Beekist honey...
325   MAHAN, P.J.   nd   A small treatise containing a few practical
directions for managing the bee and hive. Abbreviated by permission from
Rev. L.L. Langstroth's Bee-keeper's manual, by Phinehas J. Mahan, agent
for the sale of Langstroth's patent moveable comb hive. Philadelphia,
Merrihew & Thompson 32,8,4p illus   I
    1858 Baltimore, Sherwood & Co 48p   I
    "Sent anywhere for a 3 cent stamp".
    [See Bee World 50(3): 111-14 (1969)]
326   MANLOVE, J.H.   1883   The southern bee-keepers guide.
Leonard, Tex. 48p   W
    He & Allen purchased the state (Tex.) right in the adjustable bee hive,
    patented by Nelson C. Mitchell March 9, 1875 (Pat 160,695)

327   MANUM, A.E.   1879 3rd ed   Bristol hive. Bristol, Vt.·   C
   Bee supplier
328   MARTINEZ, J.J.   1885   Les abeilles. Petit manuel pratique.
New Orleans, La., Imprimerie Franco-Americaine 45p   E W
329   MASSIE, T.K.   nd   A few sane and irrefutable deductions on
the subject of beehives and practical beekeeping. Charleston, W. Va.,
Synder Print Co   C
330   MASSIE, T.K.   1903   The queen bee and the palace she should
occupy. How she may be reared, shipped and introduced into a distant
apiary with little, if any deterioration in her prolificness or injury
to her usefulness and value. Charleston, W. Va., Donnally Publ Co 75p
illus   F I M
331   MAY, A.J.   1897   Dr. H. Penoyer's improved bee-palace.
Washington, D.C., Searle Print 5p   M
332   MAYNE, W.   1955   A swarm in May. Indianapolis & N.Y., Bobbs-
Merrill Co 199p illus
   Publ England
333   MEEUSE, B.J.D.   1961   The story of pollination. N.Y., Ronald
Press Co 243p illus bibliogr
334   METCALF, M.   1862   A key to successful bee-keeping: being
a treatise on the most profitable method of managing bees, including the
author's new system of artificial swarming, whereby all watching for swarms
during the swarming season is done away with, and all loss by flight to
the woods prevented. N.Y., C.M. Saxton 96p illus   C X
   1863 2nd ed   M
   Patented a hive and sold Italian queens
   Discussion of Langstroth's claims of originality for his patented frame-
   hive on p. 28-34
335   MILLER, C.C.   1874   A book. Bein selexions frum "Song
messenger" by P. Benson, Sr. [pseud]. Vollum 1. Chicago, Geo. F. Root
& Sons 106p illus   M
   Humorous ser
336   MILLER, C.C.   [1886]   A year among the bees; being a talk
about some of the implements, plans and practices of a bee-keeper of 25
years experience, who has for 8 years made the production of honey his
exclusive business. Chicago, Amer. Bee J. 114p illus [2 different
bindings]   B C E I L M T W X
337   MILLER, C.C.   1898   Food value of honey. Why it should be
eaten. Honey cooking recipes. Denver, Colo., R.K. & J.C. Frisbee 44p
illus   X
   An ed Medina, A.I. Root Co 13p   C E W
   Used as promotional material
   An ed 11p   C M;   An ed Honey as a health food. [16p]   E
   Rev St. Thomas, Ont., Ont. Bee-keepers' Ass 19p
   See 382 & 507
338   MILLER, C.C.   1903   Forty years among the bees. Chicago,
G.W. York & Co c1902 327p illus append for 10¢ extra   B C I M T W X
   1906 2nd ed 336,[7]p append p. 322-36   C E I M T W
   M copy has notes and corrections by C.C. Miller
339   MILLER, C.C.   1911   Fifty years among the bees. Medina,
A.I. Root Co 340,[6]p, adv illus   B C E I M T W
   1915 320p   B C E L X;   1920 328p   B C I M
   Memorial ed Tribute by E.R. Root & E.F. Phillips
340   MILLER, C.C.   1917   A thousand answers to beekeeping questions
as answered by him in the columns of the Amer. Bee J. Comp by Maurice G.
DADANT. Hamilton, Amer. Bee J. 276p illus   B C F L M T W X
   1919 3rd ed E I T;   1923 4th ed   I T;   1927 5th ed   I;
   1931 6th ed   B I

341   MILLER, C.C., ROAT, J., SECOR, E. & YORK, G.W.   nd   Songs
of beedom. Chicago, George W. York & Co. 15p   E L
342   MILUM, V.G.   1964   History of our national beekeeping
organizations. Minco, Okla., Amer Beekeeping Fedn   90p
343   MINER, T.   1804   The experienced bee-keeper; or, a short
treatise on the management of bees; founded on facts and long experience:
... Litchfield, [Conn.], T. Collier 21p   W
344   MINER, T.B.   1849   The American bee keeper's manual; being
a practical treatise on the history and domestic economy of the honey-bee,
embracing a full illustration of the whole subject, with the most approved
methods of managing this insect through every branch of its culture, the
result of many years' experience. N.Y., C.M. Saxton; London, John Wiley
349p illus   B C I L M T W
    1849 2nd ed   E X;   1850   C F T;   1851 3rd ed;   1851 4th ed
    F I L;   1852   B C I M T X;   1854   E M;   1855   I;
    1857   C I W X;   1859   C E X
345   MINER, T.B.   1851   An essay on winter management of bees,
with highly important rules for feeding, and for the treatment of diseased
brood, with a plan for the certain destruction of the moth, entirely new.
Also, a statement of the great profits from bees, as kept by scientific
apiarians... Utica, [N.Y.], R. Northway & Co 24p illus   C X
346   MITCHELL, N.C.   1868   The bee-keeper's guide, being a
complete index and reference book on all practical subjects connected with
bee culture, in both common and movable-comb hives; also a summary of the
year, being a complete analysis of the whole subject. Marion, Ohio,
G.Crawford & Co. 108p illus   I
    1871 First lessons in bee culture, or bee-keeper's guide... Indianapolis,
    Ind., Indianapolis Print & Publ House 96,8p   W X
    1879 [12th ed]   I
347   [MITCHELL, N.C.]   1879   The bee-keeper's directory, a complete
guide to Mitchell's new system of bee culture. Indianapolis, Ind., Rowan
& Andrews 64p illus   I
348   MITCHELL, N.C.   [1879]   Hints to bee-keepers. Indianapolis,
Ind. 27p illus   I M
349   MITCHELL, S.H.   1871   The bee-keeper's catechism, being a
practical reference book, giving minute directions on the culture of the
honey bee... containing also the result of more than twenty year's extensive
practical experience in bee culture in the climate of Canada. Stratford,
[Ont.], Wm. Buckingham's Steam Print Office 88p   M
    Patented hive
350   MOON, A.F.   1872   Autobiography and personal recollections
and experience of Ambrose F. Moon, with forty years experience in apiculture.
Indianapolis, Ind., Indianapolis Print & Publ House 115,5p   E I
351   MOORE, R.N.   1926   Supercedure and queen rearing.   I
    Galley sheets
352   MORGAN, E.A.   nd   Bee-keeping for profit. Cumberland, Wis.,
F.F. Morgan 41p illus   C E M
353   MORGAN, H., FYLER, G.L. & HENLE, L.A.   1936   Let's cook
with California honey. Davis, Calif., Calif. Beekeepers' Ass 16p
354   MORLEY, M.W.   1899   The bee people. Chicago, A.C. McClurg
& Co 177p illus juv   E F I M W X
    1900   C I T;   1901 3rd ed   E L M T X;   1904 4th ed   C;
    1909 8th ed   C F;   1911 9th ed   I;   1913 10th ed   I X;
    1914 11th ed;   1916 12th ed   T X;   1926 18th ed   T;
    1927 19th ed;   1930 20th ed   I;   1937 rev 176p   I L W X

355   MORLEY, M.W.   1899   The honey-makers. Chicago, A.C. McClurg
& Co 424p illus   C F I L M T W
    1915   I T X
    Many references to ancient writers & customs
356   [MORRISON, W.K.]   [1907]   Beekeepers' dictionary. For use
by practical beekeepers giving all the terms now in common use by American
beekeepers. Bee-keeper's ten-cent Libr. No. 1. Medina, A.I. Root Co
22p   I M T
    1922 32p   C I M W;   An ed The Right Way Ser.   C M X
357   [MORRISON, W.K.]   [1907?]   The dovetailed hive, and Hoffman
frame. Beekeeper's ten-cent Libr. No. 27. Medina, A.I. Root Co 24p
illus   F I M
358   [MORRISON, W.K.]   1907   Handling bees. Beekeeper's ten-cent
Libr. No. 4. Medina, A.I. Root Co 15,1p illus   C F
359   [MORRISON, W.K.]   1907   The honey-flow. Beekeeper's ten-cent
Libr. No. 7. Medina, A.I. Root Co 13p   F I
360   MORRISON, W.K.   [1907]   Moving and shipping bees. Beekeeper's
ten-cent Libr. No. 29. Medina, A.I. Root Co 16p illus   C F
361   MORRISON, W.K.   [1907]   The shallow or divisible brood-
chamber hive. Beekeeper's ten-cent Libr. No. 28. Medina, A.I. Root Co
11p   F
362   MORSE, J. [& O'HARA]   1931   Following the bee line. Chicago,
Thomas S. Rockwell Co 127p illus   C F I T W X
362a   MORSE, R.A.   1967   A short history of the Empire State Honey
Producers' Association. Ithaca, N.Y., The Author 75p
363   MOUNTAIN STATES HONEY PROD. ASS.   1928   Articles of incor-
poration and by-laws as amended February 1, 1928. Boise, Idaho 26p   M
364   MUDGE, L.C.   nd   Entire new method and practical knowledge
of successful beekeeping. 8p   W
    Bee taming compound and hive attachment
365   MUNRO, J.   1936   Busy bee. N.Y., Farrar & Rinehart, Inc.
63p illus juv   W
366   MUTH, C.F.   1881   Practical hints to bee-keepers. Cincinnati
32p illus   M
    5th ed   M;   An ed   M
367   MUTH, C.F.   [1884]   Foul brood, and a new cure. [Cincinnati]
6p
368   NAILE, F.   1942   The life of Langstroth. Ed by E.F. Phillips.
Ithaca, N.Y., Cornell Univ Press 215p illus
369   [NATIONAL BEE-KEEPERS' ASS.]   [1901?]   The owner of bees is
not necessarily liable for an occasional injury done by them to others.
[Wis.] 4p;   What the courts say in relation to property in bee. Goff
vs. Kiets [Kietz]. [4p];   City bee-keeping upheld. Platteville, Wis.,
Witness Print [8p]   C
    See 180; Includes 381
370   NEAL, C.D.   1961   What is a bee. Chicago, Benefic Press
48p illus juv
371   NELSON, J.A.   1915   The embryology of the honey bee. Princeton,
N.J., Princeton Univ Press 282p illus bibliogr   B C E I L M T W X
372   NEWBERRY, M.A.   nd   Honey and how to use it. Guelph, Ont.,
Can. Honey Fed 6p recipes
373   NEWMAN, T.   1877   The honey bee. Wintering bees; how to
do it successfully. Prize essays of the Nat Bee-keepers' Ass at
Philadelphia, October 26, 1876. Chicago, _Amer. Bee J._ 28p   M
374   NEWMAN, T.G.   1878   Bee culture; or, successful management
of the apiary. Chicago, Thomas G. Newman & Son 79,1p illus   M

1879   W X
1882 3rd ed Bees and honey; or, the management of an apiary for profit
and pleasure. Chicago, Amer. Bee J. 158p illus   C E F I L M T
[1884?] [1-11],12-154,24p (p. 37,38,42-44 rev] 19 plates index misnumbered [185]-191 11p adv   E
Testimonial in Given Press adv dated 1884
[1888?] [1-11],12-154,28p (37-44 rev, 45-46 deleted) 52 plates
index misnumbered [185]-192 12 or 14p adv   C I M
A.I. Root adv announces 38th thousand of ABC of Bee-Culture but was
meant to be 37th since next known ed is 42nd thousand
1892 [1-11],12-154p (p. 45-46 deleted) 50 plates 12p adv index
misnumbered [185]-192   B L T X
[1899?] [1-11],12-154 (p. 37-44 rev) 5 plates 1p adv   C I M
G. York appears as ed in Amer. Bee J. adv
First advertised in Sept. 7, 1899 Amer. Bee J. with 160p
1911 Rev by DADANT, C.P. ...; or, first lessons in bee keeping.
190p   I L M T W X;   1915   I;   1916   T
Paper bound copy sent free to new subscribers to the Amer. Bee J.
1878 Bienen-kultur. Chicago, Tomas G. Newman & Son 84p illus   L M
[1884-1890] Amer. Bee J. Office 84,2p   B C I
375   NEWMAN, T.G.   [1878]   Honey, as food and medicine. Chicago,
Thomas G. Newman & Son 23p   M
  [1881] rev 24p;   An ed 21p   X;   [1892] George W. York & Co
23p   M X;
1898 Honey as food. Forepart by MILLER, C.C. 24p
First publ for distribution by beekeepers to "forever annihilate the
cry of "glutted market"." Amer. Bee J. 14(4): 129 (1878)
Also publ in German
Replaced by 382
376   NEWMAN, T.G.   [1882]   Bee pasturage a necessity. Chicago,
Amer. Bee J. p. 103-30 from "Bees and honey", 1882   M
377   NEWMAN, T.G.   [1882]   The production of comb and extracted
honey, preparation for the market, etc. Also, the management of bees and
honey at fairs, and expositions. Chicago, The Author p. 45-58, 1 plate,
131-40 from "Bees and honey", 1882   M
378   NEWMAN, T.G. (comp)   [1884]   Bee-keepers' convention handbook, giving rules of order for a deliberative assembly, a digest of
parliamentary law, constitution, by-laws, etc., etc. Chicago, Amer. Bee
J. 32p   E M T X
379   NEWMAN, T.G.   [1886?]   Apiary register. Chicago, Amer.
Bee J. 118p record book   C E I
380   NEWMAN, T.G.   1886   Brief history of the North American
Bee-keepers' Society, with a digest of its annual conventions from 1870
to 1884, and a full report of the proceedings of the sixteenth annual
convention, held at Detroit, Michigan, on Tues., Wed., and Thure., Dec. 8-10,
1885. Chicago, Amer. Bee J. 60p illus   B I T
381   NEWMAN, T.G.   [1889]   Bee-keeping not a nuisance. Chicago,
Nat Beekeeper's Union 8p
  Arkansas Supreme Court decision. Incorporated in 369
382   NEWMAN, T.G. & SON   1891   Honey almanac. Chicago 32p
illus recipes   I
  Rep George W. York & Co
  1899 replaced 375 for promotional purposes. See 337
383   [N.Y. STATE ASS. of BEE-KEEPERS' SOC.]   [1899]   Important
and interesting facts for farmers and fruit growers. When spraying for
injurious insects, care must be taken not to destroy those that are

beneficial.--Abstracts from bulletins and articles written by eminent
authorities on cross-fertilization and spraying. 8p
    Includes F. Benton, R.L. Taylor, G.M. Doolittle, A.J. Cook, etc.
    [See Glean. Bee Cult. 27(10): 402 (1899)]
    Incorporated in 179 & 180
384   NIVER, J.   1859   The bee-keeper's manual and guide to the
use of the patent protective bee-hive. Albany, N.Y., Weed, Parsons &
Co 48p illus   W
    Hive patented by Henry Eddy, M.D., North Bridgewater, Mass. secured
    to Niver. Testimonials for him in append p. 30-48
385   NIXON, G.   [1955]   The world of bees. N.Y., Philosophical
Libr 214p illus
    Publ London, Hutchinson
386   NIXON, M.I. (comp)   nd   Delicious "Beekist" honey recipies.
Toronto, Ont. Honey Prod Co-operative Ltd 15p
387   NOEL, M.   1886   Buz, or The life and adventures of a honey
bee. N.Y., H. Holt & Co viii,134p   X
    Publ England 1885   B M
    1888   T;   1892   C I;   1898   W
388   NORTH, O.S.   [1969]   Bee and honey patents of the world.
Arlington, Va., The Author iv,300,v-xiip illus
    Photostat copies of patents
389   NORTH AMER. BEEKEEPERS' SOC.   1872   Trans. of the North
Amer. Beekeepers' Soc. 1st Annual Session held at the City of Cleveland,
[Ohio]. Dec. 6-8, 1872 55p   M
390   N. C. BEEKEEPERS ASS.   nd   Constitution. Medina, A.I. Root
Co [8p]
391   ODELL, F.G.   nd   Modern bee-keeping. A practical hand-book
for the apiary. College View, Nebr., Rose Lawn Apiaries 40p illus
    Adv offering Abhasian, Banater, Carniolan, Red Clover & Italian queens
392   OGLESBY, R.D.   1954   Adventures in beekeeping. N.Y., Vantage
Press 52p illus glos
393   Oldfield, J.   Money for health. Toronto, Ont. Honey Prod.
Co-operative, Ltd. 16p
394   ONSTOTT, K.   1941   Beekeeping as a hobby. N.Y., Harper &
Bro viii,137p illus
395   ORDER OF BEES   1912   Ritual, order of bees (incorporated).
Greensboro, N.C., Grand Colony 16p   W
    1922   W
    Initiation dialogue based on bee lore for boys and girls of Free Masons
396   PAGDEN, J.W.   [1868?] $350 a year: How I make it by my
bees; and how others may soon do the same. Boston, Loring 45p   F X
    Publ England
397   PAISLEY, T.   1950   Betty the bee. Cleveland, Ohio, Paisley-
Bradov Co 24p illus juv
398   PARSONS, S.B.   [1861]   Circular. [Italian or Ligurian
bees]. Cited Amer. Agriculturist adv March p. 91, July p. 223 (1861);
M. Quinby "Mysteries of bee-keeping" 1869 p. 309
    [See J. of Long Island History 9(1): 21-31 (1969)]
399   PATON, J.G., CO.   1933   The importance of honey for health.
27p recipes promotional
400   PAYNE, W.   1855   The farmer's guide and bee cultivator.
Containing directions for using Payne's improved bee hive; for constructing
hive, -and hiving and dividing swarms, etc. Maine, N.Y. 7p   W
401   PEARCE, J.A.   1910   The Pearce method of bee-keeping. Grand
Rapids, Mich., The Fruit Belt Publ Co 28p illus   C F I M T X

1915 2nd ed rev   56p   C I W X
1918 3rd ed rev   The Pearce new method of bee-keeping. Joseph A.
Pearce Co 58p   F I M W X
402   PEASE, C.H.   1949   Backlot beekeeping written by a small
beekeeper with thirty years' experience, especially for the beginner who
knows nothing about bees, and the small beekeeper with limited experience,
owning from one to fifty colonies. Canaan, Conn., Canaan Print Co 110p
illus
    "The little book for little beekeepers."
403   PELLETT, F.C.   1916   Productive bee-keeping; modern methods
of production and marketing of honey. Lippincott's Farm Manuals.
Philadelphia & London, J.B. Lippincott Co xiv,302p illus
                                                    B C E F I M T W X
    1918 2nd ed rev   B C I T X;   1923 3rd ed   B E F L T W X;
    1928 4th ed rev Added: ia-ixa "Bee-keeping enterprises" consisting
    of 14 jobs or units for study   B C I T X
404   PELLETT, F.C.   1918   Practical queen rearing. Hamilton,
Amer. Bee J. 103p illus   C E M T W X
    1918 2nd ed   B E L;   3rd ed   B E F T;   1929 4th ed 104p
    B C I T W X;   5th ed repr   I;   1945 6th ed Dadant & Sons 103p
405   PELLETT, F.C.   1919   Beginner's bee book. Philadelphia &
London, J.B. Lippincott Co 179p illus glos   B C E F I L M T W X
    Interleaved, rev copy & carbon copy of manuscript at C
406   PELLETT, F.C.   1920   American honey plants. Hamilton,
Amer. Bee J. 297p illus   B C E F I L M T W X
    1923 2nd ed 392p "Publications consulted" added   B C I L M T W X
    1930 3rd ed 419p   B C E I T W X;   1947 4th ed 467p
407   PELLETT, F.C.   1923   Every step in moving bees. Every Step
Ser. No. 2. Hamilton, Amer. Bee J. [7p] illus
408   PELLETT, F.C.   1931   The romance of the hive. N.Y.,
Cincinnati, etc., Abingdon Press 203p illus   B C E F I L M W X
    Many of these stories appeared in his dep., "The busy bees" in the
    Flower Grower
409   PELLETT, F.C.   1938   History of American beekeeping. Ames,
Iowa, Collegiate Press ix,213p illus   B C F I L T W X
410   PELLETT, F.C.   1943   A living from bees. N.Y., Orange
Judd Publ Co 304p illus glos
    1944;   1945;   1946 rev 332p;   1946 332p,3p(index);
    1947 335p;   1948;   1951
411   PELLETT, F.C.   1944   The honeybee, source of the world's
most famous food. Bios Classroom Ser. No. 2. Mt. Vernon, Iowa 16p illus
412   PELLETT, F.C.   [1947]   Useful honey plants. Hamilton, Amer.
Bee J. 16p
413   PELLETT, K.L.   nd   Charles Dadant. That bee man from
Champagne. 119p typed manuscript   M X
414   PERKINS, H.   [1922]   First aid to amateur beekeepers.
Los Angeles, Calif., Miller Box Mfg Co 42p   E I M
415   PHELPS, E.W.   1853   Phelps' bee-keeper's chart; being a
brief practical treatise on the instinct, habits and management of the
honey-bee, in all its various branches. The results of many years' practical
experience, to render bee-keeping less difficult, and at the same time more
sure, profitable and pleasant than it has formerly been. N.Y., C.M. Saxton
96p illus   C
    1854   M W;   1856   E;   1858 A.O. Moore, Cover: C.M. Saxton,
    Barker & Co   C I W X

416   PHILLIPS, E.F.   1904   The habits of the honey-bee. Medina,
A.I. Root Co   26p   illus   E
     1905 24p   I M;   1906   I;
     1907 Bee-keeper's ten-cent Libr No. 17   F;   1910   I;
     1914   C I M;   1917   B E M T;
     1924 The Right Way Ser.   C I X;   Rev 13p
     1914 Las costumres de las abejas. 36p illus   C X
417   PHILLIPS, E.F.   1906   Bee-keeping for women. Medina,
A.I. Root Co 28p illus   C
     An ed Bee-keeper's ten-cent Libr No. 18. 29p   C M
418   PHILLIPS, E.F.   1915   Beekeeping; a discussion of the life
of the honeybee and of the production of honey. Rural Science Ser. N.Y.
& London, Macmillan Co xxii,457p illus   B E I L M T W X
     1916;   1917   C X;   1919   C E I T;   1920   E;
     1922   M;   1923;   1926   C;
     1928 2nd ed   xxvii,490p   B C E I L T W X;   1937   B;
     1939;   1942;   1943 (April, Dec.);   1945;   1946;   1947;
     1949;   1956
     1960 rev ECKERT, J.E. & SHAW, F.R.   See 157
419   PHILLIPS, E.F.   1936   Honey as a food and medicine. With
additions by the Bee Culture Laboratory, Dep of Agr, Washington, D.C.
Comp ROOT, E.R. Medina, A.I. Root Co 16p   C
     1941
     Promotional pamphlet. See 507
420   PHILLIPS, E.F. [& NELSON, J.A.]   1924   Beekeeping. N.Y.,
Boy Scouts of Amer 53p illus   C
     1942 62p;   1944 27p;   [1953] 29p
     1957 rev FILMER, R.F. New Brunswick, N.J. 66p
421   PHILLIPS, G.W.   1904   How to produce extracted honey. Medina,
A.I. Root Co 29p illus   C F X
     1907 Bee-keepers' ten-cent Libr No. 25   F.I;   1911 2nd ed 43p M;
     1914 3rd ed 40p   T;   1917 4th ed 34p   C I M;
     1919 5th ed 32p   C E M;
     1924 Extracting honey. The Right Way Ser 24p   C;   [1927]
     1912 Como se produce la miel extraída ó líquida. 51p illus   C I M
422   PHILLIPS, G.W.   1904   Modern queen rearing. Medina, A.I.
Root Co 31p illus   E M
     1905 3rd ed 33p errata p. a-h   C I M T;   without errata   B;
     1916 rev PRITCHARD, M.T. 24p   E I;   1924   C;
     1932 16p   I X
     1914 La cria moderna de reinas. 37p illus   C
423   [PHILLIPS, G.W.]   1914   Direcciones practicas pára los
principiantes en apicultura. Medina, A.I. Root Co 45,1p illus   C
424   PHILLIPS, M.G.   1923   Honey bees and fairy dust. Philadelphia,
Macrae Smith Co 213p illus juv   I
     1926 N.Y., D.C. Heath & Co   B C E F L M X
425   PHILLIPS, M.G.   1956   The makers of honey. N.Y., Thomas
Y. Crowell Co 163p illus bibliogr juv
426   PHILLIPS, M.G.   1967   The bee man. Life and letters of
Everett Franklin Phillips. Ithaca, N.Y., Office of Apicult, Cornell
Univ 112p mimeo
427   PHIN, J.   1884   A dictionary of practical apiculture.
Giving the correct meaning of nearly five hundred terms, according to the
usage of the best writers. Intended as a guide to the uniformity of
expression amongst bee-keepers, with numerous illustrations, notes and
practical hints. N.Y., Industrial Publ Co [xii],7-80p illus
                                                           C E I L M T W X

428   PIERCE, G.R.   1891   The winter problem in bee-keeping. An exposition of the conditions essential to success in the winter & spring management of the apiery. Oneonta, N.Y., E.C. Reynolds, Job Print   77p illus   C E M X
     An print Blairstown, Iowa   M
429   PLATT, R.   1968   Walt Disney's bees and ants. N.Y., Golden Press   176p   illus   paperback   juv
430   POLLOCK, F.L.   1917   Wilderness honey. N.Y., Century Co 325p   illus   C
     Beekeeping in Canada. Appeared serially in Youths Companion
431   POLLOCK, F.L.   1927   Honey of danger. N.Y., Chelsea House 251p   fiction   C
432   POLLOCK, F.   1935   Bitter honey. Toronto, Thomas Nelson & Sons   315p   fiction
433   PORTER, G.S.   1925   The keeper of the bees. Garden City, N.Y., Doubleday Page & Co   505p   illus   fiction
434   POTTER, T.C.   1908   Bee-keeping for sedentary folk or for professional people. Madina, A.I. Root Co   22p   illus   C F W X
     1914   23p
435   POTTER, T.C.   1911   Queenie; the autobiography of an Italian queen bee. N.Y., Moffat, Yard & Co   82p   illus   juv   B C E F I M W X
436   POUDER, W.S.   nd   The busy bees and how to manage them. Indianapolis, Ind.   24p   illus   M
437   PRATT, C.S.   nd   Buz-buz. Boston
438   [PRATT, E.L.]   1904-1906   A series of papers on apiculture by Swarthmore [pseud]. Fully illustrated hand made in Swarthmore Shop. Swarthmore, Pa., Swarthmore Apiaries
     1) 1905 2nd ed Increase. c1904 [18p]   M W
     1906 3rd ed [Medina], A.I. Root Co   28p   C E I T;   1907 4th ed E M
     2) 1904 2nd ed Baby nuclei. 47p   I T
     1905   34p   C M W
     3) 1905 Commercial queen rearing... Cell-getting by the Swarthmore labor-saving pressed-cup and interchangeable flange steel plan. 42p
                                                                          C I M W X
     2nd ed 53p   E I M T
     4) 1906 Simplified queen rearing for the honey producer. 35p
                                                                          C E I M X
     5) [1906] Cell getting.
     6) 1906 Forcing the breeding queen to lay eggs in artificial queen- cups. 25p   C E I M T W
     French and German transl of ser
439   PRATT, E.M. (ed)   1905   The honey-money stories, by Orvice Sisson, Paul Point, Albion Girard and Charles C. Miller. Chicago, George W. York & Co   64p   illus   B C E I M X
440   PURNELL, I. & WEATHERWAX, J.M.   1930   Why the bee is busy and other Rumanian fairy tales told to little Marcu by Baba Maritza. N.Y., Macmillan Co   134p   illus   juv   I
441   QUINBY, J.W.   1859   A lecture on the habits and management of the honey bee. Little Falls, N.Y., D. Ayer, Print   14p   X
442   QUINBY, M.   1853   Mysteries of bee-keeping explained: being a complete analysis of the whole subject; consisting of the natural history of bees, directions for obtaining the greatest amount of pure surplus honey with the least possible expense, remedies for losses given, and the science of "luck" fully illustrated---the result of more than twenty years' experience in extensive apiaries. N.Y., C.M. Saxton 11,376p   illus   C F I M T W X

1854   I X;   1855   I;   1857   I T;
1858 2nd ed "Recommendations" p. 377   B C M X
1858 A.O. Moore Append describing Langstroth's hive p. 377-84 (377
repeated). Append available as separate for 10¢   C
1859 8th ed 376p append p. 377-88 "Recommendation" 1p   C T W
1860 8th ed C.M. Saxton, Barker & Co; San Francisco, H.H. Bancroft &
Co 384p append p. 377-84   F L
1861   I;   1862 C.M. Saxton   B;   1864   C E M;
1918 repr 8th ed Medina, A.I. Root Co 12,382p append p. 383-93
                                                            C I M T
1865 rev Orange Judd & Co xii,17-348p   E I L M T
1866   C F I M W;   1667   I;   1878   E T;
1879 rev ROOT, L.C. Quinby's new beekeeping. 270p B C F I L M W X
[Mr. Root was Quinby's son-in-law]
1880   X;   1881   I T;   1882;   1883;
1884 rev 271p   M T W X;   1885   B I M T X;   1888   B F T;
1891   F L M W;   1900   W;   1902   X;   1905   T X;
1908   E;   1912   E I X;   1914   M;   1918   E M;
1919 Memorial to Quinby by Capt. J.E. Hetherington p. xiii-xvi T W X
Transl publ serially in Tidskrift fur Biskjotsel (Norway) beginning
with 1(3): 56 (1885)
[Univ of Md. (Natural Sc Div) used Quinby as text 1880-1881]
443   RANDOLPH, V.   1924   Life among the bees. Girard, Kans.,
Haldeman-Julius Co 64p   W
444   RANDOLPH, V.   1925   Beekeeping for profit. Girard, Kans.,
Haldeman-Julius Co 60p illus adv p. 61-64   I W X
445   RANKIN, D.F.   1936   Better beekeeping or how we made bees
pay. Huntington, Ind., Walter H. Ball Print Co 46p illus glos
bibliogr p. 43   C I M W X
446   RANSOME, H.M.   1937   The sacred bee in ancient times and
folklore. Boston & N.Y., Houghton Mifflin Co 308p illus B C I L M W X
   Publ England
447   RAUCHFASS, F.   1923   Grading comb honey. Hamilton, Amer.
Bee J. 10p illus   C L
448   REED, C.B.   nd   The value of grading bee colonies.
Bakersfield, Calif., Valley Pollination Serv 12p illus
449   REED, W.L.   [1878]   Bee-keeping of to-day. Macon, Mo.
62p Cited Amer. Bee J. 14(3): 100 (1878)
450   RENDL, G.   1933   The way of a bee. Transl Patrick Kirwan.
N.Y., Henry Holt & Co; London, Longman Green vi,168p illus B E L W X
451   REYNOLDS, O.   1946   A brief treatise on the physiology,
anatomy, etc. of the bee; together with a description of the method of
construction and using O. Reynolds' non-swarming and dividing hive.
Rochester, N.Y., Advertiser Book & Job Office 39p illus
452   RIBBANDS, C.R.   1964   Behavior and social life of honeybees. N.Y., Dover 352p
   Publ England 1953
453   RICE, G.L.   1931   Tack of the air trails, being the
experience and life work of my personal friend, Miss Tack Aphidae;
translated from the beelingula into the child-language. Los Angeles,
Calif. 187p illus juv   W
454   RICHARDSON, H.D.   1852   The hive and the honey-bee; with
plain directions for obtaining a considerable annual income from this
branch of rural economy. To which is added, an account of the diseases
of bees, with their remedies. Also, remarks as to their enemies and the
best mode of protecting the bees from their attacks. Saxton's Rural Hand
Books, 1st Ser. N.Y., C.M. Saxton & Co 72p illus   C E F W X

```
      1856  2nd ed corrected                                              M
      1857  A.O. Moore & Co [on cover]
      1858  Moore's Rural Hand Books, 1st ser.  N.Y., A.O. Moore c1852     I
      Publ Dublin 1847;  other ed
455      RICHTER, M.C.     nd      Beekeeping library of M.C. Richter.
22p  1p addendum  typewritten                                           C M
      [Richter manuscript collection at Univ. Calif., Davis]
456      RIEBEL, L.     nd      Part 1.  Hints to beginners in bee culture.
Part 2.  Advanced bee-keeping.  Chariton, Iowa  40,43,4p  index           I
      Manufacturer of artificial limbs who also invented a hive
457      RITCHIE, M.     1964      Bees.  Boston, Beacon Press  illus
paperback  juv
458      ROBERTSON, C.     nd      Flowers and insects;  lists of visitors
of four hundred and fifty-three flowers.  Carlinville, Ill.  221p         W
      Observed within 10 miles of Carlinville
459      ROBINSON, S.     [1861]      Facts for farmers.  Vol. 1.  515p
illus  bees p. 157-76, plate xii                                          I
      [No mention of L.L. Langstroth]
460      ROOD, R.N.     1962      The how and why wonder book of ants and
bees.  N.Y., Grosset & Dunlap  46p  illus  juv
461      ROOT, A.I., CO.     nd      Answers to bee questions commonly
asked by the editors of Gleanings in Bee Culture.  Medina  79p
      1914  Beekeepers' ten-cent Libr No. 32.  75p                       I M
      1922  80p      B C M;      An ed 81p      E;
      1928  The Right Way Ser  75p                                       I X
462      ROOT, A.I., CO.     nd      Bees for pleasure and profit.  Medina
40p  illus                                                              C M T
      1923      B M;      1924      C;      1928 16p      E;      Various ed  nd
463      ROOT, A.I.     nd      Better beekeeping.  A suggestion of the
bright possibilities in keeping bees by down-to-date methods with modern
equipment.  Medina  38p  illus                                          M T
      1924  32p      C
464      ROOT, A.I., CO.     nd      Eat honey.  Medina  16p                 C
465      ROOT, A.I., CO.     nd      The essentials of a good smoker.
Medina  16p  illus
466      ROOT, A.I., CO.     nd      How to build up an apiary from a
pound package of bees.  Medina  8p
      1921  ...two pounds...
467      ROOT, A.I., CO.     nd      How to sell honey.  Medina  31p
illus                                                                  C M T
      An ed 35p      C;      1950 32p
468      ROOT, A.I., CO.     nd      A look-in on beekeeping.  Medina
32p  illus
469      ROOT, A.I., CO.     nd      The use of honey in cooking.  Medina
46p  recipes                                                            C I X
      An ed 56p      M T
470      ROOT, A.I., CO.     nd      You can make comb foundation.  Medina
4p
471      ROOT, A.I., CO.     [1897?]      Foul brood.  How to know it.
Its treatment and cure.  Leafl No. 2.  Medina  8p                         C E
472      ROOT, A.I., CO.     1903      Cuba as a honey country.  Medina
31p  illus                                                                C I
      1906  30p      E
473      ROOT, A.I., CO.     1904      Books for bee-keepers, farmers,
market-gardeners, florists etc.  Medina  p. 4-21
```

474   ROOT, A.I., CO.   1904   Root's correspondence school of
apiculture. Medina 84p   M
   1905 Prospectus of the Root correspondence school of bee culture.
   c1904 11p illus   C
475   ROOT, A.I., CO.   1906   Bee-line to profit. Medina 32p
   Vignettes of staff of Glean. Bee Cult.
476   ROOT, A.I., CO.   1907   Building up colonies. The Beekeepers'
ten-cent Libr No. 6. Medina 18p illus   C
477   ROOT, A.I., CO.   1907   Handling bees. Beekeepers' ten-cent
Libr No. 4. Medina 16p illus
478   ROOT, A.I., CO.   [1907]   [Honey comb.] Medina
479   ROOT, A.I., CO.   1907   Inmates of the hive. The Beekeepers'
ten-cent Libr No. 2. Medina 15p   C
480   ROOT, A.I., CO.   1907   Spring management of bees. The
Beekeepers' ten-cent Libr No. 12. Medina 14p illus   C E I
481   ROOT, A.I., CO.   1907   Swarming. The Beekeepers' ten-cent
Libr No. 3. 7p illus
482   ROOT, A.I., CO.   [1907]   Transferring bees from box hives
or log gums to movable-frame hives. Medina 4p illus
   1923 16p   C;   1926 15p
483   ROOT, A.I., CO.   1910   The truth about sweet clover.
Medina 96p illus
   [1913]
484   ROOT, A.I., CO.   nd   Breves apuntes sobre la apicultura
móderna o el sistema Americano. Medina 47p illus   I
485   ROOT, A.I., CO.   1912   Establishing an apiary. Medina
22p illus   I
486   ROOT, A.I., CO.   1915   Directions for operating the Root
friction-drive power honey extractor. Medina
487   ROOT, A.I., CO.   [1915]   The story of the A.I. Root
Company. 50 years. Medina 40p illus
488   ROOT, A.I., CO.   1917   Power honey-extractors. Medina
32p illus   I M
   An ed 29p   M T;   An ed 15p   T
489   ROOT, A.I., CO.   1923   Beeswax. Latest and most economical
methods of rendering old combs, refining, and marketing of beeswax.
Medina 37p illus   B C I M
   [1928] Right Way Ser. 22p   C I T
490   ROOT, A.I., CO.   [1935]   Does beekeeping pay? Medina
16p illus
491   ROOT, A.I., CO.   1949   How to produce comb honey. Medina
15p illus
492   ROOT, A.I., CO.   1954   Honey an acceptable supplement for
infant feeding. Medina 15p illus
493   ROOT, A.I., CO.   1954   Beginning with package bees. 15p
illus
494   ROOT, A.I.   1877   The ABC of bee culture, a cyclopaedia of
everything pertaining to the care of the honey bee: bees, honey, hives,
implements, honey plants, &c., &c.: compiled from facts gleaned from
the experience of thousands of bee-keepers, all over our land, and
afterward verified by practical work in our own apiary. Part 1st. Medina,
   A.I. Root vi,48,4(conclusions)p illus
   An ed 1878   M;   1878 4th ed
   1878 Part 2nd p. 51-103   X
   1879 5 Parts offered separately or bound in cloth or paper 212,4p   M
   1879 [xvi],245,22p glos & index added   C E

```
1880  [viii],245,22p                                           C I T X
1881  [xvi]286,32p  Doolittle's review added                    C I
1882  [xvi],288,30p    C M;      1883 (cover 1882) xvi,288,26p     I;
1883  [xvi],288,30p    C I M X;      [1884] [xviii],308,32p    C I W X
15,000 copies of prior ed had been sold
[1886]    C I M X;      1887 (cover 1882) [xvi],308,32p    C E I M X
1888 rev  ROOT, E.R.  xvi,324,55p  biographical sketches & Miller's
review added                                                    C I T
1890  [xiv],403p    C E I T X;      1891 [xiv],404p   B C E F I M T X;
1895  [xiv],436p    B C E I M T X;
1899 rev  xii,437p    C I M T X;      1901 [x],398,365-80,417-32,[24],
461-74p    C I M T X;      1903 [x],482p    C E F I M W X;
1905  x,490p    B C E I M X
1908  ABC and XYZ of ... ROOT, A.I. & E.R.  viii,536p B C E F I M T W X
1910  viii,576p    C E I M T W X;      1913 [x],717p    C E I M T W X;
1917  [x],830p    B C E I M T W X;      1919 [x],856p    B C I T X;
1920  x,856p    C E M W X;      1923 x,960p    B C E F I M T W X;
1929  x,815p    B C E I M T W X;      1935 xiv,815p    B C E I M T W X;
1940  xiv,813p;      1945 xvi,720p;      1947;      1948;      1950 29th ed
xiv,703,xivp;      1954 30th ed;      1959 31st ad [xiii];
1962 32nd ed  glos p. 705-12;      1966 33rd ed
Transl German 1907;  French 1909 2nd ad;  Spanish 1914, 1923, 1948;
Russian 1938
495    ROOT, A.I.    1886    Merrybanks and his neighbor. Some
pictures from real life of a bee-keeper. Telling Mr. Merrybanks'
failures as well as his ultimate success after having profited by past
experience. Medina 210p illus                                C I M
496    ROOT, A.I. & ROOT, E.R.    1911    Our friends, the bees.
Washington, D.C., Judd & Detweiler  p. 675-94
    From Nat. Geographic Magazine July, 1911
497    ROOT, E.R.    nd    Bee and honey facts. Medina 4p illus    C
498    ROOT, E.R.    nd    Honey for cooking. 4p                   C
    For Redpath Chautauqua lectures
499    [ROOT, E.R.]    nd    Stereopticon lecture on bees. Medina,
A.I. Root Co  18p                                                C
500    ROOT, E.R.    1904    Beginning with bees. Medina, A.I. Root
Co  14p illus                                                    E
    1917  15p
501    ROOT, E.R.    1906    Bee diseases. How to know and how to
treat them. Medina, A.I. Root Co  18p                            C
    1908  Beekeepers' ten-cent Libr No. 13.  21p               B I M
    1918  43p    B;    1923 36p    M;
    1924  Cover title Diseases of bees. The Right Way Ser.  31p  illus
    C I X;    1935 21p    X
502    ROOT, E.R.    1907    Wintering bees. Bee-keepers' ten-cent
Libr No. 11. Medina, A.I. Root Co  23p illus                     C
    From 1908 (c1907) ABC & XYZ... "Wintering" p. 464-86
    1913  ...An exhaustive treatise of the subject covering both the
    outdoor and indoor methods.  62p                       C E I M T
    1923  ...An exhaustive treatise of the subject covering the summer
    and fall management, required for strong colonies for winter, and
    both outdoor and cellar wintering. Right Way Ser.  72p    C F I W
503    ROOT, E.R.    1908    The bee-keeper and the fruit-grower.
Why and how their interests are mutual. Medina, A.I. Root Co  4,15p    L
    [1910] 15p    F;    1911 14p;    1918 Cover title Bees and fruit.
    20p    E;    1920 29p illus    T;    1921    8 M;    1924 27p
    C;    1925    C;    1932 23p    C;    Various others 14-19p    C
```

504   ROOT, E.R.   [1911]   The buckeye bee-hive or the management
of bees in double-walled hives. Medina, A.I. Root Co 65p illus
    Other ed with 79, 87, or 93p   C I M
    1921 The buckeye hive. Why it is much the best hive for many bee-
    keepers to use; and how to manipulate the bees in it. 24p   B M T
    1924 15p   C
505   ROOT, E.R.   [1911]   Feeding and feeders. Medina, A.I. Root
Co 33p illus   C M
    [1913]
506   ROOT, E.R.   1914   Feeds and feeding. Medina, A.I. Root Co
32p illus
    1923 24p   B C I X;   An ed 33p   M
507   ROOT, E.R.   1923   Honey as a food. Medina, A.I. Root Co
[15p]
    1924 [16p];   1926;   1931;   1932
    See 337 & 419
508   ROOT, E.R.   1925   Beekeeping in a nutshell. Medina,
A.I. Root Co 32p illus   X
    1926   X
509   ROOT, H.H.   1951   Beeswax. Brooklyn, N.Y., Chemical Publ
Co v,154p illus bibliogr
510   ROSEN, E.   1969   To be a bee. Boston, Houghton 44p
illus juv
511   ROUSE, J.W.   nd   The amateur bee-keeper. A book designed
for amateurs and beginners in beekeeping. Higginsville, Mo., Leahy Mfg Co
    1894 2nd ed 63p illus;   [1895] 3rd ed 74p   F;   4th ed
    84,2p   C;   1905 5th ed 82,1p   C E M X;   1910 6th ed 87p
    B E I M T X;   1919 6th ed 81p
512   ROWE, H.G.   1922   Starting right with bees, written by a
beginner in bee-keeping...out of his own experience in needing help and
direction during his first two years of beekeeping. Read, revised and
approved by the editors of Gleanings in Bee Culture. Medina, A.I. Root
Co 128p illus   C E M X
    1927 2nd ed   C F T;   1931 3rd ed   T;
    1936 4th ed Rev ROOT, E.R. & DEYELL, M.J. 102p   C W X;
    1937 5th ed   T;   1942 6th ed 104p;   1945 7th ed;
    1947 8th ed c1946:   1947 9th ed;   1952 10th ed 100p
    1956 11th ed Ed by BARTH, W.
513   RUSSELL, F.   1967   Honeybee. N.Y., Knopf juv
514   SANDBURG, C.   1963   Honey and salt. N.Y., Harcourt, Brace
& World 11p poetry
515   SANTERRE, A.   1903   La ruche canadienne. Quebec, Imp.
Darveau 205p illus   X
516   SAURIOL, C.   [1961]   Honey is my hobby. Toronto 50p
517   SCHAFER, M.   1958   Honey's nifty fifty recipes. Calif.
Honey Advis Bd 34p illus
518   SCHAFER, M.   [1961]   The best from the west with honey.
Calif. Honey Advis Bd 30p illus recipes
519   SCHAFER, M.   [1966]   Honey: it's good every way and every
day. Whittier, Calif., Calif. Honey Advis Bd 30p recipes
520   SCHAMU, C.G.   1919   The Schamu roller entrance bottom
board and the beginners first lessons in bee-keeping. Liverpool, N.Y.
34p illus   C M
    nd 8p illus   C M
521   SCHOFIELD, A.N.   1947   Teach yourself bee-keeping.
Philadelphia, David McKay Co 151p illus
    Publ London 1943

522   SECHRIST, E.L.   1944   Honey getting. Hamilton, _Amer._ _Bee_ _J._
128p illus
   1947 2nd ed Roscoe, Calif., Earthmaster Publ 82p
523   SECHRIST, E.L.   1947   The beemaster system for hive heating.
Roscoe, Calif., Earthmaster Publ 28p illus mimeo
524   SECHRIST, E.L.   1955   Amateur beekeeping. N.Y., Devin-Adair
Co 148p illus
   1970
   Publ London 1958 Ed by BROWN, A.C.
525   SECHRIST, E.L. & McFARLAND, D.F.   1946   Scientific beekeeping. Roscoe, Calif., Earthmaster Publ 64p illus
   1948
526   SECOR, E. (ed)   1902   Bees and horticulture; some opinions
of scientific men, trained observers, experimenters, and fruit-growers
as to the value of bees in the orchard and garden. Nat Bee-keepers
Ass 14p   C
   See 179 & 383
527   [SECOR, E.   1914   The calender.]
528   SHARP, D.L.   1925   The spirit of the hive; contemplations
of a beekeeper. N.Y. & London, Harper & Bro 240p   B C E F I L M T W X
529   SHELDON, C.   1949   Lines written while under the influence
of honey mead by the King Bee! 75p
530   SHEPPARD, W.J.   1918   Beekeeping in the Kootenays British
Columbia. Nelson, B.C., Beekeepers' Ass of B.C. 15p illus
531   [SHEPPARD, W.J.]   1922   Beekeeping in the interior of
British Columbia. Nelson, B.C., B.C. Honey Prod Ass 16p illus
532   SHUTTLESWORTH, D.   1967   All kinds of bees. N.Y., Random
House 62p illus juv
533   SISSON, L.S.   1872   The bee-keepers' manual. A practical
treatise, giving minute directions in every department. Utica, N.Y.,
Curtiss & Childs 102p   C
   Includes page of testimonials for author's patent frame and bee hive
534   SKINNER, O.   1922   Bee fishing with Lurem. Gowen, Mich.
8p illus   E
535   SLATER, L.G.   1969   Hunting the wild honey bee. Olympia,
Wash., Terry Publ Co 94p
536   SMITH, C.H.   1888   A glimpse of 1888 bee culture, with
illustrations and prices of modern apiarian supplies. Pittsfield, Mass.
47p illus   E I M
537   SMITH, E.C.   1918   The hive bee; a manual of beekeeping
for Hawaii. Honolulu, Hawaiian Gazette Co 36p illus   I L M
538   SMITH, F.G.   1963   Beekeeping: a beginners' guide to
profitable honey and beeswax production. N.Y. & London, Oxford Univ
Press 129p illus
539   SMITH, H.   1970   Biography of a bee. N.Y., Putnam's &
Sons juv
540   SMITH, J.   1923   Queen rearing simplified. Medina, A.I.
Root Co 119,3p illus   B C E F L M T W X
   1949 Better queens. Washington, D.C., Judd & Detweiler, Inc
   100p illus
   "High Brow" p. 72 is Dr. Phillips
541   SMITH, J.   [1925?]   About bees. Vincennes, Ind. 20pE I M T
   1931 rev 24p   I
542   SMITH, J.   [1925]   Twenty-five years with bees. Vincennes,
Ind. 20p   C
543   SMITH, J.   1950   50 years with bees. Fort Myers, Fla.
[13p] illus

544   SMITH, J.V.C.   1831   An essay on the practicability of cultivating the honey bee, in maritime towns and cities, as a source of domestic economy and profit. Boston, Perkins & Marvin; N.Y., J. Leavitt 106p illus   C E F I L M W X
    Langstroth's copy at I: "L.L. Langstroth, my first book on bees, purchased 1835"
    [See Rep Iowa State Apiarist for 1951: 23-24 (1952)]
545   SMITH, W.D.   1921   Swarm control. Los Angeles, Calif., Clyde Broune Co 11p   E C
    Adv for Smith's top hive entrance
546   SNODGRASS, R.E.   1925   Anatomy and physiology of the honeybee. N.Y., McGraw-Hill Book Co xv,327p illus bibliogr   B C T W X
    Six? printings
    1956 Anatomy of the honey bee. Ithaca, N.Y., Comstock Publ Associates xiv,334p
547   SOUDER, D.   1807   The rural economist's assistant in the management of bees: Principally taken from the German writings of the Reverend J.L. CHRIST, First Minister of Krohnberg, and member of the royal Husbandic Society of Zelle. Lancaster, [Pa.], Print. by William Greear vi,55p illus   M
548   SOUTHERN CALIF. HONEY EXHIBIT COMM.   nd   Honey and its uses in the home. A compilation of tested receipes. [Madison, Wis.], Amer Honey Inst 31p
549   "THE SPECTATOR"   1904   My first season's experience with the honey-bee. Medina, A.I. Root Co Repr by permission of the Outlook Co N.Y.   E F
    [1905];   1911 13p;   1913   C;   1934 15p   C
550   STOCKLEY, C.   1914   Wild honey. N.Y. & London, Constable 340p fiction
551   STOCKWELL, G.A.   1891   Apiculture: the double-hive, non-swarming system. Bread & Butter Ser No. 2. Providence, R.I., Snow & Farnham 16p   W
552   STRUTT, E.   1946   Why are bees so busy? A "why story" for children who ask questions. Hollywood, Calif., Graphic Educ Prod, Inc. 16p phonograph record juv
553   STUART, F.S.   1949   City of the bees. N.Y., McGraw-Hill Book Co 243p illus fiction
    1961 Toronto, McGraw
    Publ England
554   STURGES, A.M.   1924   Practical beekeeping. Philadelphia, D. McKay xix,307p illus   B F L T
    Publ London
555   SUTHERLAND, L.   1946   The life of the queen bee. N.Y., B. Ackerman 126p illus
556   SWENSON, V.   1959   A Maxton book about bees and wasps. N.Y., Maxton Publ [29p] illus juv
557   TEALE, E.W.   1940   The golden throng. N.Y., Dodd, Mead & Co 208p illus bibliogr
    1946;   1959 A book about bees. Bloomington, Ind. Univ Press; 1961 216p
558   TEALE, E.W.   1961   Bees. Adapted by W. Rebhulen. N.Y., Columbia Record Club juv
    1967 Bees. Adventures in Nature & Science Ser. Chicago, Childrens Press 63p illus
559   TEX. HONEY PROD. ASS.   1921   The honey book. San Antonio, Tex. 32p recipes   C
    [1923?]   M

560   TEX. STATE BEEKEEPERS Womens Auxillary   1958   Honey
recipes from Texas. 55p illus
     1959 2nd ed   37,1p
561   THACHER, J.   1829   A practical treatise on the management
of bees; and the establishment of apiaries, with the best method of
destroying and preventing the depredations of the bee moth. Boston,
Marsh & Capen 162,2p   C E I L M T W X
     Inventor of hive
562   [THOMAS, I.]   1792   A complete guide for the management of
bees, through the year. By a farmer of Massachusetts. Illustrated with
a copperplate. Worcester, Mass., Print. by Isaiah Thomas & Leonard
Worcester for Isaiah Thomas 46p illus   E M W
     One of two copies at W has slightly different frontispiece
     Text evidently taken from DANIEL WILDMAN; p. 37-40 contains letter
     from Mr. CRIS JOHN on bee hunting in America; p. 40-45 a note from
     the Encyclopedia on the honeybee; p. 45-46 a similar note on the sting
     of the honeybee
     See Amer. Bee J. 73(9): 367 (1933)
563   THOMAS, J.H.   1865   The Canadian bee-keepers' guide: an
easy method of managing bees by the use of Thomas' patent moveable comb
bee hive. Toronto, Globe Steam Press 72p illus   M
     1869 6th ed   M;   1871 7th ed   T
564   THOMAS, W.   nd   Results of experiments with honey bees and
red clover. Medina, A.I. Root Co
565   THORPE, T.B.   1854   The hive of "the bee-hunter". A
repository of sketches including peculiar American characters, scenery,
and rural sports. N.Y., D. Appleton & Co 312p illus   I
566   TIBBETTS, A.B.   1952   The first book of bees. N.Y., Franklin
Watts, Inc. 69p illus juv
     1963
567   TINKER, G.L.   1890   Bee-keeping for profit, or how to get
the largest yields of comb and extracted honey. Flushing, Mich.,
Rulison's Print House 49p illus   C E I L M X
     1893 Chicago, George W. York & Co 130p   B C E I L M X
     Chap. "Bee pasturage a necessity" from NEWMAN, T.G.   374
     Adv for bee supplies & Syrio-albino bees
568   TIREMAN, L.S. & YRISARRI, E.   1945   Dumbee. Albuquerque,
N. Mex., Univ of N. Mex. Press 46p illus juv
569   TITCOMB, S.A.   1849   A brief history of the honey-bee,
with remarks upon honey, bee-hives, and the general management of bees,
designed as an accompaniment to Titcomb's patent compound bee-hive.
Farmington, Maine, J.F. Sprague, Print 28p illus   C M
570   TONSLEY, C.   1970   Honey for health. N.Y., Award Univ
Publ & Dist 125p paperbound
571   TOWNLEY, E.   1843   A practical treatise on humanity to
honey bees; or practical directions for the management of honey bees,
upon an improved and humane plan, by which the lives of bees may be
preserved, and abundance of honey of a superior quality obtained. N.Y.,
Print by Wm. S. Dorr 162p   B C F I M T X
     1848 c1843 Print by G.B. Maigne   C E F L M
     Offered rights for his patent bee hive
572   TOWNSEND, E.D.   1910   The Townsend bee book or how to make
a start in bees. Medina, A.I. Root Co 87p illus   B E I L M T X
     1912   C;   1914 82p   B C X
573   TRUE, J.M.   1936   The busy little honeybee. Chicago,
Rand McNally & Co 63p illus juv   C I L M W
     1937   X

574   TUCKER, C.M.   1872   Wings and stings: a tale for the
young. N.Y. etc., T. Nelson & Sons 108p illus juv   W
    Publ England 1855, other ed   B
575   TUPPER, E.S. & SAVERY, A.   [1872?]   Bees: their management
and culture. Publ by the Italian Bee Co. Des Moines, Iowa, G.W. Edwards,
Print 26p adv for authors' bee co   M X
576   TURNBULL, W.H.   [1958]   One hundred years of beekeeping
in British Columbia, 1858-1958. Vernon, B.C., Vernon News Ltd 137p illus
577   UNDERHILL, T.S.   1860   The leaf bee-hive and its management.
N.Y., E.O. Jenkins 24p illus
578   VANDRUFF, W.S.   1889   How to manage bees, or bee culture
for the masses. Intended to elevate the masses to a higher plane of
beekeeping. Waynesburg, Pa., Independent Job Print 203p illus   C
579   VENN, M.E.   1952   Honeybee, by Mary Adrian [pseud].
Holiday Easy Science. N.Y., Holiday House 51p illus juv
580   VINCENT, W.D.   1886   Guide to bee-keeping. Catlettsburg,
Ky., Ky. Democrat 14p   E
581   WALL, D.   1935   The tale of Bridget and the bees. Pough-
keepsie, N.Y., Artists & Writers Guild; London, Methuen 45p illus juv C
582   [WANDLE, J.T.]   1893   Bees and bee-keeping in hive and
apiary. London & N.Y., The Butterick Publ Co 40p illus   W
583   WARE, K.   1960   Let's read about bees and wasps. St. Louis,
Mo., Webster Publ Co 32p illus juv
584   WATERMAN, C.E.   1933   Apistia. Little essays on honey-
makers. Medina, The Author 138p illus bibliogr   C I M W X
585   WATSON, L.R.   1927   Controlled mating of queenbees. Hamilton,
Amer. Bee J. 50p illus bibliogr   B E F I L M T W X
    [See Amer. Bee J. 61(3): 94 (1921)]
586   WEBB, A.   1943   Beekeeping for profit and pleasure. N.Y.,
Macmillan Co 116p illus
    1945;   1947 4th print;   1948 5th print;   1952
587   WEBER, C.H.W.   1903   Formaline gas as a cure for foul
brood, or, how recent experiments prove that the foul brood germ can
positively be killed by formaline gas. At the same time saving all
combs, frames and hives to be used again by the bees. [Cincinnati], The
Author [10p] illus   E
588   WEBSTER, W.B.   nd   The book of bee-keeping. N.Y., Charles
Scribner's sons 103p illus   M
    1908 104p   I
    nd A.B.C. Guide to bee-keeping. Chicago, Frederick J. Drake & Co
    103p illus   C M X;   An ed Chicago, Alhambra Book Co   I M
    Publ London 1888 Many ed   B
589   WEEKS, J.M.   1836   A manual: or an easy method of managing
bees in the most profitable manner to their owner, with infallible rules
to prevent their destruction by the moth. Middlebury, [Vt.], Knapp &
Jewett, Print 73p   C E L M W X
    1837 2nd ed   72,2p "Supplementary remarks."   I
    1837 2nd ed   Elam R. Jewett, Print c1836 72p p. 72 misnumbered 12 C
    1838 3rd ed   Hamilton Drury, Print 93p   C X
    1839 4th ed   Brandon, Vt. Telegraph Office 96p   C E F I T X
    1840 rev   Boston, Weeks, Jordan & Co vi,128p   B C E F I M T W X
    1854 2nd new ed N.Y., C.M. Saxton Append by W.A. Flanders iv,119p C E
    1855 c1854 illus   F T;   1856   E;   [1857]
    Adv for Vt. crystal palace, or Flanders' patent bee hive
    [See Amer. Bee J. 105(10): 371 (1965)]

590   WEEKS, J.M.   1840   The bee-keepers guide to manage bees in
the Vermont bee-hive. Middlebury, Vt., Argus Office 24p   C
591   WEST, N.D.   1906   Circular and price list of spiral wire
queen-cell protectors and queen-cages. Middleburgh, N.Y. 16p illus
592   WEST, W.   1876   Guide to bee-keeping, being a series of
instructive chapters embracing the most practical information and advice
on the subject of bee culture, including extracts from the most popular
authors. Nashville, [Tenn.], C. LeRoi Print 85,2p illus   C
593   WHEELER, C.   1847   The apiarian's directory: or, practical
remarks on the economical, advantageous, easy and profitable management
of bees, to accompany and explain the New-York hive. Buffalo, N.Y.,
Charles E. Young c1836 138p   I
594   WHEELER, C.   1855   The bee-keeper's guide; or, practical
remarks on the most advantageous, easy and profitable management of bees.
To accompany and explain the eclectic hive and drone-trap. N.Y., W.L.
Burroughs 108p illus   W
595   WHITCOMBE, H.J. & DOUGLAS, J.S.   1955   Bees are my business.
N.Y., G.P. Putnam's Sons vii,245p illus
    Publ London 1956
596   WHITEHEAD, S.B.   1951   Honeybees and their management.
Ed & rev SHAW, F.R. N.Y., D. Van Nostrand Co 169p illus bibliogr
    Publ London 1946
597   WHITMAN, O.H.   1947   First steps in bees. How to form
package bees, queen's nuclei, extraction and transferring. Davenport,
Iowa   I
598   WIESTLING, J.S.   1819   Der vollstendige Bienen-Wärter,
oder Nützliche Anweisungen zur Bienen-Zucht... Harrisberg, Pa., The
Author [2,3],4-50p illus   [At Amer. Antiquarian Soc.]
599   WILDER, J.J.   nd   Chunk honey production. 15p illus   C
600   WILDER, J.J.   1908   Southern bee culture. [Cordele, Ga.]
143p illus   C E I M T X
601   WILDER, J.J.   1927   Wilder's system of beekeeping. Waycross,
Ga. 93p illus   B C E I L M T X
602   WILHELM, M.   [1965]   Honey cook book. Erick, Okla.,
Wilhelm Honey Farm 126p
603   WILKES, A.H.   1914   A scientific study on swarming and how
to prevent it. Birmingham, [Ala.], Four Oaks
604   WILKIN, R.   1868   Hand-book in bee-culture. Pittsburgh,
[Pa.], W.S. Haven & Co 95,2p illus   C M
    1871 96,2p   C I M X
    Dealer in hives
605   WILLIAMS, G.W.   nd   The food value of honey. Redkey, Ind.,
United Honey Prod 16p illus   C
    An ed Honey, its food value.
606   WIS. STATE BEEKEEPERS' ASS.   1920   Second directory and
revised constitution. Madison, Wis. 32p   M
607   WOLF, C.W.   1858   **Apis mellifica**; or, the poison of the
honey-bee, considered as a therapeutic agent. Philadelphia, William
Radde 80p   C M W
608   WOOD, J.G.   [1863]   Bees; their habits, management and
treatment. London & N.Y., G. Routledge & Sons new ed   E T
    Publ London 1853 Many ed   B
609   WOODMAN, A.G., CO.   nd   Honey cookery. 16p recipes
610   WORKING, D.W.   1902   Something about the bee industry in
Colorado. A souvenir of the Denver meeting of the Nat. Bee-Keepers'
Ass. Sept. 3-5, 1902. Denver, Colo., App-Sutherland Eng Co [37,11p]
illus adv   C E F I L M X

611   WRIGHT, A.T.   1873   Universal bee hive. Pat No. 85,716,
Jan. 5, 1869. Chicago, J.F. Daniels 19p   W
612   YORK, G.W.   [1904?]   The story of the American Bee Journal.
14p illus   C
613   ZINN, H.J.   1945   Our Martha. [Dunmore, Pa.], Henry J.
Zinn & Co 152p illus
    Collection of humorous letters

DOMINION OF CANADA PUBLICATIONS

Dominion publications have been produced by the Apiculture
Service (originally Bee Division and then Apiculture Division) of
the Department of Agriculture with headquarters at the Central
Experimental Farm in Ottawa since 1893, and associated Research
Stations located in the various provinces since 1887. In 1959,
The Department of Agricultural Research Service fused with the
Experimental Farms Service and the titles were changed from Api-
culture Division to Apiculture Service and Experimental Farms to
Research Stations.

The federal organization is currently concentrating on
research which is published in the Canadian Entomologist and
other Canadian, United States and British journals. Practical
applications of the research are disseminated by the Information
Section of the Canadian Department of Agriculture and the
provincial Departments of Agriculture. Apicultural research is
now centered in the Entomology Section of the C.D.A. Research
Branch.

Titles followed by an asterisk [*] are also published in
French.

C1   L'ARRIVEE, J.C.M. & GEIGER, J.E.   1964   Package bees for
profit.° Pub 1227 10p illus
C2   BRAUN, E.   1936   Package bees in Manitoba; eleven years
experimental results 1921-31. Pub 522, Farm Bull 11 18p illus
C3   BRAUN, E.   1940   Two methods of wintering bees for the
prairie provinces. Pub 689, Circ 160 4p illus
. 1955 rev Braun, E. & GEIGER, J.E.   Comparison of methods of
   wintering. 8p
C4   BRAUN, ·E.   1945   Dividing over-wintered colonies for
increased honey production. Pub 774, Farm Bull 130 13p illus
C5   BRAUN, E. & JAMIESON, C.A.   1955   Tubular heat exchange
equipment for processing honey. 4p 2 plates
C6   BRAUN, E. & PANKIW, P.   1953   Moving honeybee colonies.
5p illus
C7   CHERCUITTE, U. & FOREST, B.   1950   Comte rendu de vingt-
cinq ans de recherches apicoles (1923-47) a'la Station Experimentale
Federale de Sainte-Anne-de-la-Pocatière. 34p illus
C8   ENGLISH, J.K.   1939   Canadian honey in the United Kingdom.
United Kingdom Ser 2 39p illus
C9   FIXTER, J.   1903   Bee keeping. Evidence of Mr. John
Fixter [Apiarist, Central Exp Farm] before the select standing committee
on agriculture and colonization, 1903. 21p
   1905 Management of bees-care of combs and honey. [16p]
C10   GEIGER, J.E. & L'ARRIVEE, J.C.M.   nd   Package bees, their
introduction and care. 14p illus
C11   GOODERHAM, C.B.   [1921]   Bee diseases. Exhibition Circ 105
4p
C12   GOODERHAM, C.B.   1930   Package bees and how to install
them. Pam 107 N S 8p illus
   1936 rev Pub 507, Farm Bull 8 11p
C13   GOODERHAM, C.B.   1939   Preparing bees for winter. Pub 674,
Circ 151 4p illus
C14   GOODERHAM, C.B.   1940   Bees. Spring management. Spec Pam 2
4p
C15   GOODERHAM, C.B.   1940   Package bees. Spec Pam 3 4p
C16   [GOODERHAM, C.B.]   1946   Beekeeping. Small Holdings Ser 9
135p illus
   Test questions after each topic
C17   GOODERHAM, C.B. & HEENEY, M.L.   1934   Honey and some of the
ways it may be used. Pam 161 N S 15p recipes
   1936 1st rev Pub 501, Household Bull 2 20p
C18   HEWITT, C.G.   1912   The honey bee, a guide to apiculture in
Canada.° Bull 69 45p illus
C19   INLAND REVENUE DEP LABORATORY   1897   Honey. Bull 47 21p
   1903 Bull 90 15p;   1906 Bull 122 11p;   1908 Bull 148 17p
   Results of honey analysis
C20   INLAND REVENUE DEP LABORATORY   1908   Strained honey.
Bull 145 27p
   1910 Bull 217 19p
C21   JAMIESON, C.A.   1939   Care of bees and equipment. Pub 685,
Circ 157 4p illus
C22   LOCKHEAD, A.G. & HERON, D.A.   1929   Microbiological studies
of honey. Bull 116 N S 47p illus
C23   SLADEN, F.W.L.   1912   Bees and how to keep them.* Bull 26
2nd Ser 56p illus

```
     1916;      1923 rev  GOODERHAM, C.B.      Bull 33 N S  60p;      1926;
     1929;      1934 rev  Bull 169  62p;      1945   Pub 578, Farm Bull 37
     60p;      1947
C24      SLADEN, F.W.L.      1915      Beekeeping in Canada.*  Exhibition
Circ 18  4p  illus
     1920 rev
C25      SLADEN, F.W.L.      1916      Facts about honey.*  Exhibition
Circ 51 rev  3p
     1920 rev  4p
C26      SLADEN, F.W.L.      1920      Wintering bees in Canada.*  Bull 43
2nd Ser  12p  illus
     1922 rev  GOODERHAM, C.B.  Pam 22 N S  13p:      1926 rev  Bull 74 N S
     31p;      1940  Pub 681, Farm Bull 89  27p
C27      WHALLEY, M.E.      1934      Dietetic and medicinal value of honey.
Nat Res Council  30p  bibliogr
```

PROVINCIAL PUBLICATIONS

In each province the office of a Provincial Apiarist is
the center for the Extension Service and work of the Inspectors.
There is a close association between this office and the bee-
keepers' organizations. The agency listed for each province
is the current source of information and has published most of
the provincial publications. Publications from other governmental
agencies within the province are given complete citations.

Annual reports, reprints from serial publications and
inspection reports have been omitted unless they include a broad
treatment of beekeeping within the province.

ALBERTA   Department of Agriculture, Edmonton, Alberta

1   Le MAISTRE, W.G.   1940   Hints to beginners in beekeeping.
Circ 23   4p   illus
2   Le MAISTRE, W.G.   1941   Beekeeping for beginners in Alberta.
Bull 35   45p   illus
   1948   40p;   1953 rev   40p
3   Le MAISTRE, W.G.   1947   Wintering bees in Alberta.   Bull 58 rev
18p   illus
4   Le MAISTRE, W.G.   1948   Care of package bees in Alberta.
Bull 77   10p   illus

BRITISH COLUMBIA   Department of Agriculture, Victoria, B.C.

1   CORNER, J.   1954   Beehive construction for beginners.   Apiary
Circ 10   [7p]   illus
2   [HARRIS, L.]   [1911]   ... Foul brood among bees. Bull 31   15p
3   HARRIS, L. & TODD, F.D.   1913   Apiculture in British Columbia.
Bull 42   63p
4   SHEPPARD, W.J.   [1928]   Directions for using the "top
entrance".   2p   illus;   The middle entrance hive.   2p   illus;
Final report on "top entrance" trials, 1928.   2p   mimeo
5   SHEPPARD, W.J., FINLAY, A.W. & ROBERTS, J.F.   1923   Bee culture
in British Columbia.   Bull 92   60p
   1924 2nd ed   66p;   1925 3rd ed   70p;   1928 4th ed   73p;
   1938 5th ed rev   Finlay, A.W.   56p
6   TODD, F.D.   1911   Guide to bee-keeping in British Columbia.
Bull 30   52p   illus
7   TODD, F.D.   1918   Honey production in British Columbia season
of 1917.   Circ 19   4p

MANITOBA   Department of Agriculture and Immigration, Winnipeg, Man.

1   FLOYD, L.T. & MITCHENER, A.V.   1925   Beekeeping.   Ext Bull 78
31p   illus
   1926 rev
2   MARTIN, E.C.   1947   Diseases and pests of bees. Pub 208
16p   illus
3   MITCHENER, A.V.   1931   Package bees.   Ext Bull 97   19p   illus
4   MITCHENER, A.V.   1957   A history of beekeeping in Manitoba to
1956.   Pub 290   11p
5   MUCKLE, R.M.   1915   Beekeeping in Manitoba.   Ext Bull 18
24p   illus
6   ROBERTSON, D.R.   1953   Package bees for Manitoba.   Pub 261
8p   illus
7   ROBERTSON, D.R.   1953   Management of honey bees in Manitoba.
Pub 263   7p   illus
8   ROBERTSON, D.R.   1954   Diseases of honey bees.   Pub 268
[8p]   illus
9   ROBERTSON, D.R.   1961   Honey bee management in Manitoba.   15p
mimeo
10   ROBERTSON, D.R. & SMITH, D.L.   1964   The prevention and control
of honey bee diseases with drugs and antibiotics.   5p   mimeo
11   SMITH, D.L. & McRORY, D.G.   1968   Wintering bees in Manitoba.
6p

NOVA SCOTIA   Department of Agriculture, Truro, N.S.

1   KARMO, E.A.   1962   Beekeeping in Nova Scotia.   Part I.
Apiary management.   31p

NOVA SCOTIA

2   KARMO, E.A.   1967   The use of drugs and antibiotics in
controlling bee diseases. Mimeo Circ 76 6p
3   KARMO, E.A.   1971   Wintering of bee colonies in Nova Scotia.
Circ 82 11p illus
4   KARMO, E.A. & VICKERY, V.R.   [1959]   The place of honey
bees in orchard pollination. Mimeo Circ 67 11p illus
5   PAYNE, H.G.   1942   Beekeeping in Nova Scotia. Bull 6 79p
illus
    1947 rev 87p

ONTARIO   University of Guelph, Guelph, Ont.

1   BRADY, G.D. & BURKE, P.W.   1954   Steam boilers for the honey
house. Circ 224 10p illus
2   BURKE, P.W.   1952   Classifying and grading of honey. Circ 150
10p illus
3   BURKE, P.W.   1954   Preparing honey and beeswax for competition.
Circ 215 9p
4   BURKE, P.W.   1955   Beekeeper's honey house. Circ 225 6p
illus
    1962
5   BURKE, P.W.   1961   Specialized beekeeping practices.* Pub 25
20p illus [Includes the following circ rev by Burke]
    Burke, P.W. & ADIE, A. Packing bees in asphalt paper. Circ 132
    TOWNSEND, G.F. & Adie, A. Package bees. Circ 124;   Moving bees.
    Circ 130;   Handling the cappings from honey combs. Circ 171
    Townsend, G.F. & Burke, P.W. Feeding bees. Circ 131
6   DEP AGR   1954   Honey and how to use it. Toronto Circ 227 5p
7   DEP AGR   1910   Uses of vegetables, fruits and honey.
Toronto Bull 184 24p recipes
    1915 rev
8   DYCE, E.J. & MARTIN, E.C.   1934   Bee diseases. Bull 377
24p illus
9   HARRISON, F.C.   1900   Foul brood of bees. Toronto, Dep Agr
Bull 112 32p illus
10   MILLEN, F.E.   1920   Transferring of bees. Toronto, Dep Agr
Circ 27 12p illus
11   PETTIT, M.   1908   Bee-keeping in Ontario. Toronto, Dep Agr
Bull 166 8p
    1910 Bull 182 7p;   1911 Bull 191 3p
    Rep of honey prospects in various counties
12   PETTIT, M.   1911   Bee diseases in Ontario. Toronto, Dep
Agr Bull 190 11p
    1912 Bull 197 16p illus;   1913 rev Bull 213 19p;   1915 rev;
    1920 rev MILLEN, F.E. Bull 276 23p;   1926 rev Bull 317 22p
13   PETTIT, M.   1915   Natural swarming of bees and how to
prevent it. Toronto, Dep Agr Bull 233 15p illus
14   PETTIT, M.   1915   Some results of co-operative experiments
on races of bees to determine their power to resist European foul brood.
Toronto, Dep Agr 8p
15   PETTIT, M.   1917   The wintering of bees in Ontario. Toronto,
Dep Agr Bull 256 24p illus
16   SMITH, M.V.   1952   Honeybees for pollination. Circ 133 7p
illus
17   SMITH, M.V.   1963   The OAC pollen trap. 2p mimeo

ONTARIO

18   TOWNSEND, G.F.   1950   Bee diseases and pests of the apiary.
Bull 429 26p
    1955 suppl SMITH, M.V. 2p illus
    1965 rev Townsend, G.F., BURKE, P.W. & Smith, M.V. 36p illus
19   TOWNSEND, G.F.   1952   Beekeeping in northern Ontario.
Circ 128 4p illus
    1964 rev Townsend, G.F. & ROWLAND, M.J. 5p
20   TOWNSEND, G.F.   1952   General information on beekeeping in
Ontario. Circ 153 4p bibliogr
    1957 rev 5p;   1958 rev;   1962 rev
21   TOWNSEND, G.F.   1961   Preparation of honey for market.
Pub 544 24p illus [Combines information previously published as follows]
    Townsend, G.F. & ADIE, A. Melting honey for repacking. Circ 121;
    Straining honey. Circ 207;   The OAC continuous flow honey pasteurizer.
    Circ 216;   Honey extracting on a small scale. Circ 217;   The OAC
    pressure strainer. Circ 218;   Uniform granulation of honey by
    continuous flow. Circ 285
    Townsend, G.F. & BURKE, P.W. Removal of moisture from honey. Circ 123
    1965 rev 29p
22   TOWNSEND, G.F. (ed)   [1962]   Apiculture research in the
Western Hemisphere. 16p 1p index
23   TOWNSEND, G.F. & BURKE, P.W.   1952   Beekeeping in Ontario for
honey production and pollination. Bull 490 53p illus
    1962 rev 41p
24   TOWNSEND, G.F. & SMITH, M.V.   1953   Removal of bees from
buildings and trees. Circ 157 3p illus
    1954 rev 4p;   1960;   1963 rev Smith, M.V. 5p
25   TOWNSEND, G.F. & SMITH, M.V.   1953   Caring for bees in schools.
Circ 169 6p illus
    1963 rev

QUEBEC   Department of Agriculture, Quebec, P. of Quebec

1   ANON   1920   Aviculture et apiculture. 14p
2   BELAND, H. & VAILLANCOURT, C.   1917   Methode de cultiver les
abeilles dans la province de Québec. Bull 36 68p illus
3   BOSSÉ, L.   1942   Précis d'apiculture. Bull 155 45p illus
4   CANTIN, C.   1950   Valeur et utilisation du miel. Bull 166
35p French-Canadian recipes
    1961
5   LAJOIE-VAILLANCOURT, B.   1920   Use of honey and of maple
sugar in cooking.* Bull 68 16p recipes
6   PLOURDE, H.J.   1967   Manuel d'apiculture. Pub 297 59p
7   VAILLANCOURT, C.   1919   Le rucher Quebecois. Bull 62 78p
    1920 87p illus;   1923 L'apiculteur pratique. 94p;   1926
    Cover title: Gardez des abeilles. 112p
8   VAILLANCOURT, C.   1924   La loque; comment la reconnaitre
et la traiter. Bull 85 16p illus

SASKATCHEWAN   Department of Agriculture, Prince Albert, Sask.

1   ANON   1958   Exhibiting and judging honey and beeswax. 14p
illus
2   ARNOTT, J.H. & BLAND, S.E.   1954   Beekeeping in Saskatchewan.
Regina 92p 1 plate illus
    [See SASK 7]
3   BLAND, S.E.   1960   Pollination. 5p

SASKATCHEWAN

4   BLAND, S.E.   [1963]   Money in comb honey. Regina 18p illus
5   GRENFALL, J.H.   1922   Possibilities of beekeeping in
Saskatchewan.
6   PATTERSON, C.J.   1924   Beekeeping, a new industry for
Saskatchewan. Saskatoon, Univ Sask. Bull 21 6p
7   PUGH, R.M.   1948   Beekeeping in Saskatchewan. Regina
39p illus
    [See SASK 2]

UNITED STATES FEDERAL PUBLICATIONS

Unless indicated to the contrary, U.S. government publications cited below have eminated from the Department of Agriculture established in 1862. The many revisions in titles of bureaus, divisions, services, and administrations have not been indicated. The majority of apiculture bulletins have come from (1) Division of Entomology, (2) Bureau of Entomology, and (3) Bureau of Entomology and Plant Quarantine [See Weber, G.A., The Bureau of Entomology. Serv. Monogr. 60, Brookings Inst., Washington, D.C. (1930)]. Current publications are from the Entomology Research Division of the Agricultural Research Service. Information about federal agencies is contained in Beekeeping in the United States, Agriculture Handbook No. 335 (1967), and L.O. Howard, A History of Applied Entomology, Smithsonian Misc. Collect. Vol. 84 (1930).

With a few exceptions, annual reports, mimeographed flyers of transitory value, folders, and reprints from serial publications have been omitted. Articles and data from federal serial publications may be found in:

Annual Report, Commissioner of Patents 1837-61. [See Hitz & Hawes (F64) p. 85-87 for contents from 1845 of the Agricultural Division of the Patent Office begun in 1842]
Annual Report of the Dep. Agr. & Bur. Entomol. 1862- [See Hitz & Hawes p. 85-87 for contents to 1923]
Monthly Reports 1867-76
Insect Life (Bur. Entomol.) 1888-94
Yearbook of Agriculture 1894- [See Hitz & Hawes p. 100-04 for contents to 1928]
Weekly Newsletter 1913-21
Journal of Agr. Res. 1913-
Official Record 1922-28

Market and crop statistics for honey and wax have appeared serially in:

Crop Reporter 1913
Monthly Crop Reporter 1913-22
Semi Monthly Reports - Honey 1917-22 superseded by Honey Market News
Food Surveys 1918-19
Market Reporter 1920-21
Weather, Crops and Markets 1922-23
Crops and Markets 1924-26

Publications are arranged alphabetically by author or federal agency if other than Division or Bureau of Entomology, or Entomology Research Division. When no author is given for the latter publications, they are listed alphabetically by title at the beginning.

F1   1918   The agricultural situation for 1918. Part IV. Honey.
Circ 87 8p
F2   1969   Beekeeping for beginners. Home & Garden Bull 158   12p
illus
     1971 rev
F3   1967   Beekeeping in the United States. Agr Handbook 335
147p illus ;   1971 rev
F4   1962   Care and handling of honey. CA 33-22-62 2p
F5   1942   The dependence of agriculture on the beekeeping industry--
A review. E-584 39p annotated bibliogr
     1946 rev 42p
F6   1953   Honey...some ways to use it. Home & Garden Bull 37   16p
     1959;   1961;   1963 rev
     Supersedes Leafl 113 See F161
F7   1967   Identifying bee diseases in the apiary. Agr Inform
Bull 313 19p illus
     Supersedes Circ 392 See F35
F8   1932   Information about bee culture. E-276 5p mimeo
     1932-53 various rev;   1955-66 rev ARS-33-10-1,2,3
     Issued earlier as Keep bees better. See F9
F9   1917   Keep bees better. Mimeo Circ 31
     An ed Mimeo 172 3p
     Superseded by E-276 See F8
F10   1932   List of dealers in beekeeping supplies, package bees and
queens. mimeo
     1936 Circ E-297 10p: 1939 6p: 1940 10p: 1943;
     1950 15p
F11   1911   Miscellaneous papers on apiculture. Bull 75 123p illus
     Seven papers by BROWNE, C.A., GATES, B.N., WHITE, G.F., & PHILLIPS,
     E.F. Also published separately [listed under author]
F12   1965   Plans and dimensions for a 10-frame bee hive. ARS
CA 33-24 1p
     Earlier 1p plans nd or no.
F13   [1970?]   Pollination rental. Leafl 549
F14   1961   Proprionic anhydride for repelling bees from honey supers.
Correspondence Aid 33-16 2p
F15   1967   Protecting honey bees from pesticides. Leafl 544 6p
illus
F16   1907   Report of the meeting of inspectors of apiaries,
San Antonio, Tex., Nov. 12, 1906. Bull 70 79p
     Contains papers by FRANCE, N.E., PARKER, F.A., PHILLIPS, E.F., STEWART,
     C., & WHITE, G.F. and a record of discussion.
F17   nd   Standard labels or stamps for United States honey grades.
2p illus
F18   1932   United States standards for [grades of] extracted honey.
3p
     1943:   1951 4th issue 6p
F19   1933   United States standards for grades of comb honey. 16p
illus
     1957 repr
     Rev of standards in Circ 24 See F128
F20   ADAMS, R.L. & TODD, F.E.   1939   Cost of producing extracted
honey in California. Tech Bull 656 34p illus
F21   AGR. ADJUSTMENT ADMIN.   1934   Marketing agreement [No. 43]
and license [No. 54] for shippers of package bees and queens produced in
the United States. 16p

F22   ANDERSON, E.D.   1969   An appraisal of the beekeeping industry. ARS 42-150 38p illus
F23   BAIN, H.M.   1945   Organizing honey marketing cooperatives in wartime. Misc Rep 79 iv,28p illus
F24   BENTON, F.   1895   The honey bee: a manual of instruction in apiculture. Bul1 1 MS 118p illus
     1896 rev;   1899
     Trynsl Tokyo [1900?] 174p
F25   BENTON, F.   1897   Bee keeping. Farm Bull 59 32p illus
     1905 rev 47p
     Transl Mexico 1898 59p
F26   BISSON, C.S., VANSELL, G.H. & DYE, W.B.   1940   Investigations on the physical and chemical properties of beeswax. Tech Bull 716 24p illus
F27   BOHART, G.E.   1961   Research on legume pollination. Bull 431 p. 51-59
F28   BRICE, B.A., TURNER, A., WHITE, J.W., SOUTHERLAND, F.L., FENN, L.S. & BOSTWICK, E.P.   1951   Permanent glass color standards for extracted honey. AIC-307 5p illus
F29   BROWNE, C.A.   1907   Methods of honey testing for bee keepers. p. 16-18 from Bull 75 See F11
F30   BROWNE, C.A. & YOUNG, W.J.   1908   Chemical analyses and composition of American honeys including a microscopical study of honey pollen. Bur Chem Bull 110 93p illus
F31   BRYAN, A.H., GIVEN, A. & SHERWOOD, S.   1912   Chemical analysis and composition of imported honey from Cuba, Mexico, and Haiti. Bull 154 21p illus
     Chemical literature on honey from 1907-11 p. 17-21
F32   BUR. HUMAN NUTRITION & HOME ECON.   1948   School lunch recipes using honey. 7p mimeo
F33   BUR. HUMAN NUTRITION & HOME ECON.   1952   Quantity recipes using honey. 13p mimeo
F34   BURNSIDE, C.E.   1930   Fungus diseases of the honeybee. Tech Bull 149 42p illus
F35   BURNSIDE, C.E. & STURTEVANT, A.P.   1936   Diagnosing bee diseases in the apiary. Circ 392 34p illus
     1949 rev 31p
     Replaced by Agr Inform Bull 313 See F7
F36   BURNSIDE, C.E. & VANSELL, G.H.   1936   Plant poisoning of bees. E-398 10p 2 plates
F37   CARR, E.G.   1916   A survey of beekeeping in North Carolina. Bull 489 16p
F38   CASTEEL, D.B.   1912   The behavior of the honey bee in pollen collecting. Bull 121 36p illus
F39   CASTEEL, D.B.   1912   The manipulation of the wax scales of the honey bee. Circ 161 13p illus
F40   CLAFFEY, J.B., TURKAT, V.A. & ESKEW, R.K.   1961   Estimated cost for producing dried honey commercially. ARS 73-33 16p
F41   CLEMENTS, F.E. & LONG, F.L.   1923   Experimental pollination; an outline of the ecology of flowers and insects. Carnegie Inst Pub 386 vii,274p illus bibliogr
F42   COOK, A.J.   1892   Report of apicultural experiments in 1891. In Reports of observations and experiments in the practical work of the division made under the direction of the entomologist. Bull 26 p. 83-92 See F68

F43   DEMUTH, G.S.   1912   Comb honey. Farm Bull 503 47p illus
    1917;   1919 rev Commercial comb-honey production. Farm Bull 1039
    40p;   1932 rev;   1939 33p
F44   DEMUTH, G.S.   1921   Swarm control. Farm Bull 1198 47p
illus
    1921 (Oct.) 30p;   1927;   1931 28p;   1939
    Transl French 1928 61p
F45   DEP. AGR. STATISTICS   1967   Honey and beeswax. Bull 388
67p
F46   DEP. COMMERCE   1934   Containers for extracted honey.
Simplified practice recommendation. R 156-34 9p
    1937 rev R156-37 14p
    1941 rev Extracted-honey packages. R 156-41 10p
F47   DETROY, B.F.   1966   A cloth strainer for honey conditioning
systems. Prod Res Rep 90 6p
F48   DETROY, B.F.   1967   Valve for filling honey cans. ARS 42-130
4p
F49   FARRAR, C.L.   1938   New recommendations for the installation
of package bees, using a spray and direct-release method. E-427 7p
2 plates
F50   FARRAR, C.L.   1944   Productive management of honeybee
colonies in the northern states. Circ 702 28p illus
    1950
F51   FARRAR, C.L.   1946   Two-queen colony management. E-693
14p illus
    1958 rev ...for production of honey. ARS 33-48 11p
F52   FARRAR, C.L.   1963   Large-cage design for insect and plant
research. ARS 33-77 10p illus
F53   FARRAR, C.L., & SCHAEFER, C.W.   1939   A preliminary report
on the influence of stock on supersedure, or loss of queen bees. E-473
13,2p
F54   FELDSTEIN, L.   1911   The refractive index of beeswax.
Bur Chem Circ 86 3p
F55   FRIEDMAN, H.   1955   The honey-guides. Smithsonian Inst,
U.S. Nat Mus Bull 208 292p illus bibliogr
F56   GATES, B.N.   1908   Bee diseases in Massachusetts. p. 23-32
from Bull 75 See F11, Mass 4
F57   GATES, B.N.   1909   Beekeeping in Massachusetts. p. 81-109
illus from Bull 75 See F11, Mass 5
F58   GATES, B.N.   1914   The temperature of the bee colony.
Bull 96 29p illus
F59   GOLDSBOROUGH, G.H.   1958   Expanding markets for honey.
Agr Mktg Serv 11p
F60   HAMBLETON, J.I.   1925   The effect of weather upon the
change in weight of a colony of bees during the honey flow. Bull 1339
52p illus bibliogr
F61   HAMBLETON, J.I.   1933   The treatment of American foulbrood.
Farm Bull 1713 13p illus
F62   HAMBLETON, J.I.   1945   The indispensable honeybee.
Smithsonian Rep for 1945 p. 293-304 illus
    1962 rev The honey bee. Pub 4494 p. 465-78 illus
F63   HARP, E.R.   1966   A simplified pollen trap for use on
colonies of honey bees. ARS 33-111 4p illus
F64   HITZ, V.E. & HAWES, I.L. (comp)   1930   List of publications
on apiculture contained in the U.S. Department of Agriculture Library and
in part those contained in the Library of Congress. U.S. Dep Agr Libr,
Bibliogr Contribution 21 218p mimeo

F65   HUNT, C.L. & ATWATER, H.W.   1915   Honey and its uses in the home. Farm Bull 653 26p illus recipes
    1918;   1919;   1922 rev;   1924 21p;   1926;   1928; 1930

F66   JONES, S.A.   1915   Honeybees: wintering, yields, imports and exports of honey. Bull 325 12p illus

F67   JONES, S.A.   1918   Honeybees and honey production in the United States. Bull 685 61p illus

F68   LARRABEE, J.H.   1893   Experiments in apiculture 1892. In Reports of observations and experiments in the practical work of the Division.   Bull 30 p. 57-64 See F42

F69   LUNDIE, A.E.   1925   The flight activities of the honeybee. Bull 1328 37p illus

F70   McCRAY, A.H. & WHITE, G.F.   1918   The diagnosis of bee diseases by laboratory methods. Bull 671 15p illus

F71   McINDOO, N.E.   1916   The sense organs on the mouth-parts of the honey bee. Smithsonian Misc Collect V 65(14) 55p illus

F72   McINDOO, N.E. & DEMUTH, G.S.   1926   Effects on honeybees of spraying fruit trees with arsenicals. Dep Bull 1364 31p illus

F73   MACKENSEN, O. & ROBERTS, W.C.   1948   A manual for the artificial insemination of queen bees. ET-250 33p illus bibliogr

F74   MACKENSEN, O. & TUCKER, K.W.   1970   Instrumental insemination of queen bees. Agr Handbook 390 28p illus bibliogr

F75   Merrill, J.S.   1957-1959   Selected references.
    1957 rev Composition and therapeutic properties of honey.   9p
    1959 Antibiotics and drugs for the control of bee diseases.   4p
    1959 Antibiotics and drugs for acarine disease.   2p
    1959 Possible industrial uses for honey.   2p

F76   MICHAEL, A.S.   1954   American foulbrood of honey bees - how to control it.   Farm Bull 2074 12p illus
    Superseded Farm Bull 1713 See F61

F77   MILNER, R.D. & DEMUTH, G.S.   1921   Heat production of honeybees in winter. Bull 988 18p illus

F78   MOELLER, F.E.   1961   The relationship between colony populations and honey production as affected by honey bee stock lines. Prod Res Rep 55 20p illus

F79   NELSON, J.A., STURTEVANT, A.P. & LINEBURG, B.   1924   Growth and feeding of honeybee larvae. Dep Bull 1222 37p illus

F80   NOLAN, W.J.   1925   The brood-rearing cycle of the honeybee. Dep Bull 1349 55p illus

F81   NOLAN, W.J.   1932   Breeding the honeybee under controlled conditions. Tech Bull 326 49p illus

F82   NOLAN, W.J.   1932   The development of package-bee colonies. Tech Bull 309 44p illus bibliogr

F83   NOLAN, W.J.   1937   Bee breeding.   Yearbook Separate 1604 p. 1396-418 illus

F84   OERTEL, E.   1939   Honey and pollen plants of the U.S. Circ 554 64p illus

F85   OERTEL, E.   1960   Honeybees in production of white clover seed in the southern states. ARS-33-60 8p

F86   OTANES, F.Q.   1927   Honey bees and how to raise them. Philippine I., Bur Agr Circ 199 45p illus
    In Spanish 29p

F87   OWENS, C.D.   1958   Automatic controls for honey extractors. ARS 42-15 8p

F88   OWENS, C.D.   1962   A scale for weighing beehives. ARS-42-72 10p illus

F89   OWENS, C.D. & DETROY, B.F.   1965   Selecting and operating
beekeeping equipment. Farm Bull 2204 24p illus
F90   · OWENS, C.D. & FARRAR, C.L.   1967   Electric heating of honey
bee hives. Tech Bull 1377 24p illus
F91   OWENS, C.D. & McGREGOR, S.E.   1964   Shade and water for the
honeybee colony. Leafl 530 8p illus
F92   PEDERSEN, M.W., TODD, F.E. & LIEBERMAN, F.V.   1950   A portable
field cage. ET-289 9p illus
F93   PHILLIPS, E.F.   1905   The rearing of queen bees. Bull 55
32p illus
   Transl Italian 1908 38p; Serbian 1911 49p
F94   PHILLIPS, E.F.   1906   The brood diseases of bees. Circ 79
5p
F95   PHILLIPS, E.F.   1907   Production and care of extracted
honey. p. 1-15 from Bull 75 See F11
F96   PHILLIPS, E.F.   1907   Wax moths and American foul brood.
p. 19-22 from Bull 75 See F11
F97   PHILLIPS, E.F.   1909   A brief survey of Hawaiian bee-keeping.
p. 43-58 illus from Bull 75 See F11
F98   PHILLIPS, E.F.   1909   The status of apiculture in the U.S.
p. 59-80 from Bull 75 See F11
F99   PHILLIPS, E.F.   1910   Bees. Farm Bull 397 44p illus
   1911 rev Farm Bull 447 48p;   1917 44p;   1919;   [1921];
   1922;   1924 rev 37p;   1930
F100   PHILLIPS, E.F.   1911   The occurrence of bee diseases in the
United States. (Prelim Rep) Circ 138 25p
F101   PHILLIPS, E.F.   1911   The treatment of bee diseases.
Farm Bull 442 22p illus
   1916 20p
F102   PHILLIPS, E.F.   1914   Porto Rican beekeeping. P.R. Exp
Sta Bull 15 24p illus
   1915 In Spanish 28p
F103   PHILLIPS, E.F.   1917-1919   Memorandum to field men. I-LIII
[107p] mimeo
   Declaration of war by U.S. in 1917 created a demand for increased
   production of honey. These instructions from Phillips to 17 extension
   men include speeches, letters, extracts and articles by himself and
   others on a wide range of topics.
F104   PHILLIPS, E.F.   1918   The control of European foulbrood.
Farm Bull 975 16p illus
   1920 repr Feb., March, Nov.;   1921
F105   PHILLIPS, E.F.   1920   Control of American foulbrood. Farm
Bull 1084 15p illus
   1921;   1922
F106   PHILLIPS, E.F.   1922   The insulating value of commercial
double-walled beehives. Dep Circ 222 10p
F107   PHILLIPS, E.F.   1922   The occurrence of diseases of adult
bees. Dep Circ 218 16p illus
F108   PHILLIPS, E.F.   1923   The occurrence of diseases of adult
bees, II. Dep Circ 287 34p
F109   PHILLIPS, E.F.   1925   The bee-louse, Braula coeca, in the
U.S. Dep Circ 334 12p
F110   PHILLIPS, E.F. & DEMUTH, G.S.   1914   The temperature of the
honeybee cluster in winter. Bull 93 16p illus
F111   PHILLIPS, E.F. & DEMUTH, G.S.   1915   Outdoor wintering of
bees. Farm Bull 695 12p
   1921;   1922;   1922 (Aug.) rev;   1924;   1929

F112   PHILLIPS, E.F. & DEMUTH, G.S.   1918   The preparation of
bees for outdoor wintering. Farm Bull 1012 20p illus
     An copy 24p;   1921;   1922
F113   PHILLIPS, E.F. & DEMUTH, G.S.   1918   Wintering bees in
cellars. Farm Bull 1014   21p   illus
     An copy 24p;   1922;   1925;   1928;   1931
F114   PHILLIPS, E.F. & DEMUTH, G.S.   1919-20   Government short
course on beekeeping, 13 lectures, comp by CALKINS, R.E. mimeo
F115   PHILLIPS, E.F. & DEMUTH, G.S.   1922   Beekeeping in the
buckwheat region. Farm Bull 1216 26p illus
     1927 21p;   1930
F116   PHILLIPS, E.F. & DEMUTH, G.S.   1922   Beekeeping in the
clover region. Farm Bull 1215 27p illus
     1922 (Aug.) rev;   1924 22p
F117   PHILLIPS, E.F. & DEMUTH, G.S.   1922   Beekeeping in the
tulip-tree region. Farm Bull 1222 25p illus
     1928
F118   PHILLIPS, E.F. & WHITE, G.F.   1912   Historical notes on
the causes of bee diseases. Bull 98 96p
F119   PROD. & MKTG. ADMIN.   1950   Condensation of honey industry
- PMA Conference. 89,10p mimeo
     Discussion of price support for honey
F120   PURDON, R.L.   1926   The honey trade in the U.S. Dep
Commerce 10p mimeo
     1928 rev 13p;   1929
F121   QUICK, W.J.   1919   Bee keeping. To the disabled soldiers,
sailors & marines. Opportunity Monogr, Vocational Rehabilitation Ser 37.
Fed Board for Vocational Educ 30p illus
F122   RASMUSSEN, M.P.   1932   Economic aspects of the marketing
of honey. Prelim Rep. Bur Agr Econ 139p
F123   RODRIGUEZ, D.A.   1932   Problemas apicolas de Puerto Rico.
San Juan, P.I. Dep Agr y Comercio Circ 99 22p
F124   SCHAEFER, C.W. & FARRAR, C.L.   1941   The use of pollen
traps and pollen supplements in developing honeybee colonies. E-531
7p, 5 fig
     1946 rev
F125   SCHMITZ, G.H.   1943   Beeswax. Misc Commodities Div,
Office of Imports 13p mimeo
F126   SECHRIST, E.L.   1918   Transferring bees to modern hives.
Farm Bull 961 14p illus
     1922;   1923;   1924;   1925;   1929;   1930;
     1932 rev HAMBLETON, J.I. 9p;   1939;   1946
F127   SECHRIST, E.L.   1925   The color grading of honey.
Dep Circ 364 7p illus
F128   SECHRIST, E.L.   1927   United States standards for honey.
Dep Circ 410 32p illus
     1927 (Dec.) rev U.S. grades, color standards, and packing requirement
     for honey. Circ 24
     1933 rev MARVIN, G.E. & SPANGLER, R.L. 28p;   1943 rev 4p mimeo
     See F129, F18, F19
F129   [SECHRIST, E.L.]   1928   Requirements for the more important
grades of honey. Suppl Circ 24 [See F128] 1p
     1933 rev
F130   SECHRIST, E.L. & KIFER, R.S.   1928   Preliminary report on
apiary organization and honey production in the intermountain states in
1928 18p illus mimeo

F131   SEIN, F., Jr.   1923   Las abejas en los cafetales. P.R.
Insular Exp Sta Circ 79 6p illus
F132   SNODGRASS, R.E.   1910   The anatomy of the honey bee. Tech
Ser 18 162p illus
F133   SNODGRASS, R.E.   1942   The skeleto-muscular mechanisms
of the honey bee. Smithsonian Misc Collect Vol 103(2) 120p illus
F134   STURTEVANT, A.P.   1920   A study of the behavior of bees in
colonies affected by European foul brood. Bull 804 28p illus
F135   STURTEVANT, A.P.   1926   The sterilization of American
foulbrood combs. Dep Circ 284 28p illus
F136   STURTEVANT, A.P., KNOWLTON, G.F., HITCHCOCK, J.D., VANSELL, G.H.,
HOLST, E.C., & NYE, W.P.   1941   A report of investigations of the
extent and causes of heavy losses of adult honeybees in Utah. E-545
18p 1 plate
F137   TODD, F.E. & BISHOP, R.K.   1941   The role of pollen in the
economy of the hive. E-536 7p 2plates
    1946 rev
F138   TOWER, W.V.   1911   Bee keeping in Porto Rico. P.R. Agr Exp
Sta Circ 13 32p illus
    1913 31p
    In Spanish 1911 39p
F139   VANSELL, G.H.   1942   Factors affecting the usefulness of
honeybees in pollination. Circ 650 31p illus
    1945;   1947 rev
F140   VANSELL, G.H.   1951   Use of honey bees in alfalfa seed
production. Circ 876 11p illus
    1952
F141   VANSELL, G.H. & BISSON, C.S.   1940   Brief presentation of
the characteristics, contaminants, processing and uses of beeswax.
Circ E-495 11p 3 plates
F142   VANSELL, G.H. & TODD, F.E.   1948   Bee-gathered pollen in
various localities on the Pacific coast. E-749 9p illus
F143   VANSELL, G.H., & WATKINS, W.G.   1936   The distribution of
California buckeye in the South-Central Sierra Nevada counties, in
relation to honey production. E-397 5p 11 maps
F144   WASHBURN, R.S. & MARVIN, G.E.   1935   Organization and
management of apiaries producing extracted honey in the white clover
region. Tech Bull 481 43p illus
F145   WHITCOMB, W.   [1931]   Recommendations for shipping cages
for bees. E-287 9p 1p amendment 1 plate
F146   WHITCOMB, W.   1935   The shipping of package bees. E-363
7p 1 plate
F147   WHITCOMB, W.   1936   The wax moth and its control. Circ
386 13p illus
    1942 rev 10p
F148   WHITCOMB, W.   1965   Controlling the greater wax moth...
a pest of honeycomb. Farm Bull 2217 12p illus
    1967 rev
F149   WHITE, G.F.   1906   The bacteria of the apiary, with
special reference to bee diseases. Tech Ser 14 50p
F150   WHITE, G.F.   1907   The cause of American foul brood.
Circ 94 4p
F151   WHITE, G.F.   1908   The relation of the etiology of bee
diseases to the treatment. p. 33-42 from Bull 75 See F11
F152   WHITE, G.F.   1912   The cause of European foul brood.
Circ 157 15p illus

F153   WHITE, G.F.   1913   Sacbrood, a disease of bees.   Circ 169
5p
F154   WHITE, G.F.   1914   Destruction of germs of infectious
bee diseases by heating. Bull 92   8p   illus
F155   WHITE, G.F.   1917   Sacbrood. Bull 431   54p   4 plates
F156   WHITE, G.F.   1919   Nosema-disease. Bull 780   59p   4 plates
F157   WHITE, G.F.   1920   American foulbrood. Bull 809   46p
8 plates   illus
F158   WHITE, G.F.   1920   European foulbrood. Bull 810   39p
8 plates
F159   WHITE, J.W., Jr., REITHOF, M.L., SUBERS, M.H. & KUSHNIR, I.
1962   Composition of American honeys. Tech Bull 1261 124p   illus
F160   WHITE, J.W., Jr. & WALTON, G.P.   1950   Flavor modification
of low-grade honey. AIC-272   8p   5 tables
F161   WHITEMAN, E.F. & YEATMEN, F.W.   1936   Honey and some of
its uses. Leafl 113   8p
    1940
    Replaced by F6
F162   WILEY, H.W.   1892   Food and food adulterants. Part 6.
Sugar, molasses and syrup, confections, honey and beeswax. Div Chem
Bull 13   ix,p. 633-874   bibliogr
F163   WOODROW, A.W.   1938   Equipment and procedure for longevity
studies with caged honeybees. E-438   6p   1 plate   mimeo
F164   WOODROW, A.W.   1941   Some effects of temperature, relative
humidity, confinement, and type of food on queen bees in mailing cages.
E-529   13p   1 plate
F165   WOODROW, A.W.   1948   Tests with DDT on honey bees in small
cages. E-763   14p
F166   VAN ZWALUWENBURG, R.H. & VIDAL, R.   1918   Rearing queen
bees in Porto Rico. P.R. Circ 16   12p   illus
    Spanish transl

Addendum

F167   OWENS, C.D.   1971   The thermology of wintering honey bee
colonies. Tech Bull 1429   32p   illus

STATE PUBLICATIONS

For each state, the agency listed has published most of
the state bulletins and is the current source of state information.
In the majority of cases, this is the state college of agriculture at a land grant college (recently redesignated as state
university) which has an agricultural experiment station and/or
extension service under joint auspices of the state and the U.S.
Department of Agriculture. County Extension Agents (local
representatives of State Colleges of Agriculture and the U.S.
Department of Agriculture) are assuming a greater role in
providing information to individuals seeking assistance while
State Extension personnel are undertaking greater responsibility
for basic research with a less prominent role in providing
information at the "retail" level.

Complete citations are given for publications from other
governmental agencies within the state. Inspection reports,
State laws, and annual reports are omitted unless they include
a broad treatment of beekeeping within the state.

ALABAMA   Auburn University, Auburn, Ala. 36830

1   BASKIN, C.C. & BLAKE, G.H.   1966   Beekeeping in Alabama.
Circ P-64 35p illus
2   RUFFIN, W.A.   1937   Beekeeping in Alabama. Circ 155 11p
illus
    1942 rev

ALASKA   University of Alaska, College, Alaska 99735

1   WASHBURN, R.H.   1956   Beekeeping equipment. Entomol. Mimeo 4
2p

ARIZONA   University of Arizona, Tucson, Ariz. 85721

1   AGR. EXP. STA.   nd   [Folders on insecticides & bees]
    Citrus blossoms, bees and insecticides. Folder 87 4 col
    Melon pollination, bees and insecticides. Folder 90 4 col
    Bees and insecticides. Folder 91 5 col
    Cotton, bees and insecticides. Folder 92 5 col
2   BEEKEEPERS STATE OF ARIZONA   1924   A honey book. Phoenix,
Ariz. Comm Agr & Hort 15p recipes
3   BUTLER, G.D., TODD, F.E. & TUTTLE, D.M.   1956   Alfalfa
pollination in Arizona. Rep 130 10p
4   FORBES, R.H.   1903   Bee products in Arizona. Timely Hints for
Farm 48 8p
    1905 Bull 51
5   McGREGOR, S.E., CASTER, A.B. & FROST, M.H.   1947   Honeybee
losses as related to crop dusting with arsenicals. Tech Bull 114 27p
6   McGREGOR, S.E. & VORHIES, C.T.   1947   Beekeeping near cotton
fields dusted with DDT. Bull 207 19p illus
7   RONEY, J.N.   1946   Beekeeping for beginners. Circ 126 12p
illus
8   RONEY, J.N.   1956   Beekeeping in Arizona. Circ 238 15p illus
9   VINSON, A.E.   1906   Honey vinegar. Timely Hints for Farm 60
8p
    1908 Bull 57

ARKANSAS   University of Arkansas, Fayetteville, Ark. 72701

1   BAERG, W.J.   1920   Beekeeping in Arkansas. Bull 170 32p
illus
    1934 10p;   1937 rev Circ 326 18p;   1945 17p;
    1952 rev Baerg, W.J., THOMPSON, V.C. & BARNES, G. 20p
    1965 rev Barnes, G. & WARREN, L.D. 23p
2   POLLINATION CONFERENCE   1970   The indispensable pollinators.
A report of the Ninth Pollination Conference, Hot Springs, Ark.
Oct. 12-15, 1970 Ext Serv MP 127 233p illus
3   THOMPSON, V.C.   1960   Nectar flow and pollen yield in south-
western Arkansas 1945-1951. Rep Ser 94 38p illus

CALIFORNIA   University of California, Davis Campus, Calif. 95616

1   ADAMS, R.L. & TODD, F.E.   1933   Cost of producing queen and
package bees in California. Giannini Found Mimeo Rep 30 14p
2   ANDERSON, L.D. & ATKINS, E.L., Jr.   1967   Toxicity of
pesticides and other agricultural chemicals to honey bees - field study.
AXT-251 7p
    1971 rev 8p
    Toxicity... - lab studies. M-16

CALIFORNIA

3   BAUER, F.W.   1960   Honey marketing. Bull 776 71p
4   COLEMAN, G.A.   1917   Beekeeping for the fruit-grower and
small rancher, or amateur. Circ 185 11p illus
5   DALY, H.V.   1964   Skeleto-muscular morphogenesis of the
thorax and wings of the honey bee, _Apis_ _mellifera_ (Hymenoptera, apidae).
Berkeley, Univ. Calif. Pub in Entomol Ser 39 77p illus
6   ECKERT, J.E.   1936   Beekeeping in California. Circ 100 71p
illus
   1941 rev 86p;   1947 rev 95p;   1954 rev A handbook on...
   Man 15 89p;   1960 rev 96p
7   ECKERT, J.E.   1943   The home apiary. 4p
   1944 rev
8   ECKERT, J.E.   1954-59   [Series of leaflets]
   1959 rev Introduction to beekeeping. Leafl 14 6 col illus
   1954 Diseases and enemies of the honeybee. Leafl 29 11 col illus
      1959 rev
   1954 The honeybee colony. Leafl 31 7 col illus
   1954 Honeybees in crop pollination. Leafl 32 7 col illus
   1958 The observation hive. Leafl 102 7 col illus
   1959 Queen bees and their care. Leafl 109 7 col illus
9   ECKERT, J.E. & ALLINGER, H.W.   1939   Physical and chemical
properties of California honeys. Bull 631 27p
10   GARY, N.E. & STANGER, W.   1968   An observation bee hive.
OSA 208 2p illus
11   HENDRICKSON, A.H.   1916   The common honey bee as an agent in
prune pollination. Bull 274 6p
   1918 2nd rep. Bull 291 22p illus
   1922 Further experiments in plum pollination. Bull 352 19p
12   NICKELS, L.J.   1911   How to make an observation hive. Circ 63
13p illus
13   PARSONS, R.A., BLACK, J.H. & STANGER, W.   1969   Beehive loader
for a pickup truck. OSA 242 2p illus
14   [PHILLIPS, E.F., DEMUTH, G.S. & MENDELSON, M.H.]   nd   Lectures
of apiculture course at Davis.
   [Copies at C and Moir Library, Scotland]
15   PHILP, G.L. & VANSELL, G.H.   1932   Pollination of deciduous
fruits by bees. Circ 62 27p illus
   1944 rev 26p
15a REED, A.D.   1971   An economic analysis of the California bee
industry. MA-29 18p illus
16   RICHTER, M.C.   1911   Honey plants of California. Bull 217
64p illus
17   STANGER, W.   1965   Beginning in beekeeping. Leafl 183 5 col
18   STANGER, W.   1967   Honey-bees - agriculture's tool. OSA 188
2p
19   STANGER, W.   1967   How to remove bees from buildings. OSA 161
2p illus
20   STANGER, W.   1968   Why and how honey bees should be protected.
AXT-268 6p
21   STANGER, W. & PARSONS, R.A.   1969   Bee hive - California plan.
OSA-217 2p illus
22   STANGER, W. & THORP, R.W.   1967   Honey bees in alfalfa
pollination. AXT-228 4p illus
   1970 rev
   1967 Honey bees in almond pollination. OSA-196 2p

CALIFORNIA

23   STANGER, W., THORP, R.W. & FOOTE, L.   1968   Honey bee
pollination in California. AXT-252 10p illus
24   TODD, F.E.   1928   Bee diseases in California. Sacramento,
Calif. Dep of Agr Monthly Bull, Part 1 17(3) 14p illus
      1929 Part II 18(1) 23p illus
25   VANSELL, G.H.   1926   American foulbrood and its control.
Circ 307 12p illus
26   VANSELL, G.H.   1926   Buckeye poisoning of the honey bee.
Circ 301 12p illus
27   VANSELL, G.H.   1929   Beekeeping for the beginner in California.
Circ 36 52p illus
28   VANSELL, G.H.   1931   Nectar and pollen plants of California.
Bull 517 60p illus
      1941 rev Vansell, G.H. & ECKERT, J.E.   76p
29   VANSELL, G.H. & DEONG, E.R.   1925   A survey of beekeeping
in California. The honeybee as a pollinizer. Circ 297 22p illus
30   VANSELL, G.H., WATKINS, W.G. & HOSBROOK, L.F.   1940   The
distribution of California buckeye in the Sierra Nevada in relation to
honey production. 22p illus
      [Includes nectar, pollen & honeydew plants of the Sierra Nevada]
31   VOORHIES, E.C., TODD, F.E. & GALBRAITH, J.K.   1933   Economic
aspects of the bee industry. Bull 555 117p illus
32   VOORHIES, E.C., TODD, F.E. & GALBRAITH, J.K.   1933   Honey
marketing in California. Bull 554 31p

COLORADO   Colorado State University, Fort Collins, Colo. 80521

1   ALFORD, F.C.   1908   Extraction of beeswax. Bull 129 14p
2   BOGGS, N.   1922   Beekeeping in Colorado. Circ 37 15p illus
      1923 rev Circ 41 18p
3   BROSE, C.M.   1888   Report of experiments in apiary. Bull 5
4p
4   FOSTER, W.   1912   Loss from foul brood and poor management.
Circ 3 10p
5   FOSTER, W.   1914   Colorado apiary inspection. Circ 11 8p
6   GILLETTE, C.P.   1900   Apiary experiments. Bull 54 28p illus
7   MOFFETT, J.O.   1951   Fundamentals of bee culture. Bull 418-A
52p illus
      1965 rev WILSON, W.T.   Beekeeping in Colorado. 57p
8   MOFFETT, J.O.   1954   The use of antibiotics in controlling
European foulbrood of honeybees. Tech Bull 53 32p illus
9   RICHMOND, R.G.   1924   Wintering bees in Colorado. Circ 45
12p illus
      1926
10   RICHMOND, R.G.   1932   Red-clover pollination by honeybees
in Colorado. Bull 391 22p illus
11   RICHMOND, R.G.   1934   Introduction to beekeeping. Circ 62
28p illus
12   SACKETT, W.G.   1919   Honey as a carrier of intestinal
diseases. Bull 252 18p
13   WILSON, W.T., MOFFETT, J.O. & HARRINGTON, H.D.   1958 Nectar
and pollen plants of Colorado. Bull 503-S 72p illus

CONNECTICUT   University of Connecticut, Storrs, Conn. 06268

1   BAKER, E.L.   1947   Beekeeping spring care of colonies.
Folder 11 7 col

CONNECTICUT

2   COLEY, H.W.   nd   Diseases of bees: their detection and
treatment. New Haven, Exp Sta Bull of Immediate Inform 8   4p
    1918
3   CRANDALL, L.B.   1920   The fall feeding of bees. Bull 21
11p illus
    1932 rev Bull 166   7p
4   CRANDALL, L.B.   1921   Spring management of bees. Bull 33
11p
    1929 rev Bull 142
5   CRANDALL, L.B.   1922   Honey as a food. Bull 57   8p
    1939 rev Bull 283
6   CRANDALL, L.B.   1923   Fall management of bees. Bull 66   7p
    1930 rev Bull 147   9p
7   CRANDALL, L.B.   1926   Beekeeping a beginner's handbook.
Bull 98   12p
    1932 rev Bull 161;   1938 rev Bull 258   7p;   1942 rev Bull 337
    8p
8   YATES, A.W.   1918   Beekeeping for Connecticut. New Haven,
Exp Sta Bull 205 [22p] illus bibliogr

DELAWARE   University of Delaware, Newark, Del. 19711

1   AMOS, J.M.   1937   Roadside marketing of honey in Delaware.
Circ 32   6p illus

FLORIDA   University of Florida, Gainesville, Fla. 32601

1   ARNETT, C.M.   1954   Honey in the home. Bull 159
2   ARNOLD, L.E.   1954   Some honey plants of Florida. Bull 548
47p illus
3   DEP. AGR.   1943   Florida bees and honey. Tallahassee Bull 117
53p
4   DEP. AGR.   1950   The honey bee's service to agriculture.
Honey bees in Florida. Tallahassee Bull 135   76p illus
5   HAYNIE, J.D.   1949   Five-deep brood frame hive. Circ 90
4p illus
    1962 rev Circ 90A
6   HAYNIE, J.D. & MURPHEY, M.   1948   Lessons in Florida beekeeping
for 4-H boys and girls. 21p mimeo
    1953 rev;   1961
7   HAYNIE, J.D. & MURPHEY, M.   1952   Beginning beekeeping.
Bull 151 31p illus
    1959 rev Bull 171 32p
8   HAYNIE, J.D. & SKINNER, T.C.   1964   Plans for beekeeping
equipment and structures. Circ 272 23p
9   HOFFER, F.J. & HAMILTON, H.G.   1962   Marketing Florida honey.
Agr Econ Mimeo Rep 62-10 51p
10   HORTON, W. & THURSBY, I.S.   1933   Florida honey and its
hundred uses. Tallahassee, Dep Agr Spec Ser 66   55p illus
    1938;   1946 rev 79p with WHITFIELD, J.A. Tupelo honey. Bull 98
    10p;   1951 rev 87p with WILDER, J.J. Beekeeping in Florida.
    Bull 5 25p [FLA 13];   1958 rev 80,29p
11   KILLINGER, G.B. & HAYNIE, J.D.   1952   Honeybees and other
factors in Florida's legume program. Bull 497 14p illus
12   MORSE, R.A.   1956   Florida beekeeping. Gainesville, State
Plant Bd   Vol 2 Bull 10 113p illus

FLORIDA

13   WILDER, J.J.   1928   Beekeeping in Florida. Tallahassee,
Dep Agr Bull 5 N S   24p   illus
    1938 rev 34p;   1940 rev 43p
    [See FLA 10]

GEORGIA   Coastal Plain Experiment Station, Tifton, Ga.   31794

1   GIRARDEAU, J.H.   1954   Reseeding crimson clover as a major
honey plant in South Georgia. (Prelim rep) Mimeo Ser N S 1 8p
2   GIRARDEAU, J.H.   1958   The mutual value of crimson clover and
honey bees for seed and honey production in South Georgia. Mimeo Ser 63
23p
3   MURPHEY, M., Jr.   1941   Beekeeping. Atlanta, Dep Entomol
Bull 82 15p illus

HAWAII   University of Hawaii, Honolulu, Hawaii   96822

1   ECKERT, J.E.   1951   . Rehabilitation of the beekeeping industry
in Hawaii. IRAC Grant 19 Final Rep 29p illus
2   ECKERT, J.E. & BESS, H.A.   1952   Fundamentals of beekeeping
in Hawaii. Bull 55 59p illus
3   VAN DINE, D.L. & THOMPSON, A.R.   1908   Hawaiian honeys.
Bull 17 21p illus

IDAHO   University of Idaho, Moscow, Idaho   83843

1   DEP. ENTOMOL.   1956   . Protect your pollinators. Bull 258   4p

ILLINOIS   University of Illinois, Urbana, Ill.   61822

1   DADANT, C.P. (comp)   1913   State law on bee diseases description
and treatment. State Bee-keepers' Ass. Apiary Inspection Bull 1   15p
illus
    1914 KILDOW, A.L. Springfield, Dep Agr Apiary Inspection Bull 2   22p
    illus;   1918 Circ 261 11p;   1919;   1923;   1927;
    1940 rev KILLION, C.E. Bee diseases cause and treatment. 22p;
    1943;   1958 rev 26p
2   JAYCOX, E.R.   1969   Beekeeping in Illinois. Circ 1000   132p
3   JAYCOX, E.R.   1969-1971   (Reports from Horticulture)
    1971 rev Evaluating honey bee colonies for pollination. 3p illus
    Fruit Growing No 20
    1971 rev Pollination of fresh vegetable and canning crops. 3p illus
    Vegetable Growing No 20
    1971 rev Pollen inserts for apple pollination. 3p Fruit Growing No 22
    1969 Making and using pollen inserts. 2p illus Fruit Growing No 23
4   JAYCOX, E.R.   1969-1971   (Extension circulars)
    1969;   1971 rev Pesticides and honey bees. Circ 940   5 col
    1969 Destroying bees and wasps. Circ 1011   5 col
    1971 Pollination of legume seed in Illinois. Circ 1039   5 col
5   JAYCOX, E.R.   [1971]   Construction details 10-frame bee hive.
1p illus
6   JAYCOX, E.R.   1971   (Extension sheets)
    How to move bees. H-669 2p illus
    Managing honey bees for pollination. H-670 6p
    Rearing wax moth larvae. H-671 2p
7   MILUM, V.G.   1939   A 4-H club manual for the honey production
project. 20p mimeo illus
    Illinois 4-H record book for honey production project. 19p mimeo

ILLINOIS

8   MILUM, V.G.   1943   Illinois honey and pollen plants.   11p mimeo
    1951 rev   12p;   1957 rev   13p
9   MILUM, V.G.   1944   Honey production.   Circ 566   52p   illus
    1945 rev;   1947;   1952 rev;   1960 rev
10   MOYER, S.I.   1942   How to use honey.   Circ 528   16p   recipes
11   MOYER, S.I.   1942   Use honey.   HEE-3064   14p   mimeo
12   TRACY, P.H.   1932   How to make honey-cream.   Bull 387   12p illus   recipe
13   TRACY, P.H., RUEHE, H.A. & SANMANN, F.P.   1930   Use of honey in ice-cream manufacture.   Bull 345   16p

INDIANA   Purdue University, Lafayette, Ind.   47907

1   BALDWIN, E.G.   1919   Swarm control for Indiana.   Bull 83   8p illus
2   BALDWIN, E.G.   1919   Wintering bees in Indiana.   Bull 85   8p illus
3   [BALDWIN, C.H. & KINDIG, B.F.]   1913   Circular of information to beekeepers. The foul brood diseases. Indianapolis, State Entomol. 16p   illus
    1914 rev   Kindig, B.F.   Circular of information for beekeepers.   No 2 [43p]
4   DAVIS, J.J.   1925   Beekeeping in relation to agriculture. Bull 129   8p   illus
5   DEP CONSERVATION   1933   The brood diseases of bees in Indiana and their control. Indianapolis, Pub 127   19p
6   DIETZ, H.F.   1925   Pollination and the honey bee.   Indianapolis, Dep Conservation Pub 52   20p   illus
7   JORDAN, R.   1925   Honey its use in the home.   Circ 121   12p illus recipes
8   JORDAN, R.   1942   Stretching your sugar supply with honey, molasses, corn sirup, maple sirup.   Bull 284   8p   recipes
9   JORDAN, R.   1953   Honey recipes that satisfy.   Mimeo   HE-171 7p
10   MONTGOMERY, B.E.   1936   Recognition and control of honeybee brood diseases.   Bull 212   8p   illus
    1938 rev;   1950 rev   Honeybee brood diseases.   12p
11   PRICE, W.A.   1920   Bees and their relation to arsenical sprays at blossoming time.   Bull 247   15p   illus
12   PRICE, W.A.   1929   Beekeeping in Indiana.   Bull 158   24p
    1936 rev   MONTGOMERY, B.E.   20p   illus;   1940 2nd rev;
    1950 3rd rev   28p
13   TROOP, J. & PRICE, W.A.   1917   Bees for the farmer.   Circ 61 20p   illus bibliogr
14   YOST, C.O.   1924   Brood and adult bee diseases in Indiana. Indianapolis, Dep Conservation   Pub 43   16p
    1927 rev   23p

IOWA   Iowa State University, Ames, Iowa   50010

1   GROUT, R.A.   1937   The influence of size of brood cell upon the size and variability of the honeybee (_Apis mellifera_ L.).   Res Bull 218   p. 260-80   illus
2   KENOYER, L.A.   1917   The weather and honey production. Bull 169   26p   illus

IOWA

3   PADDOCK, F.B.   1914   Radio-correspondence course.   Bk-53-77
124p mimeo
4   PADDOCK, F.B.   1922   Beginning with bees. Bull 98 23p illus
5   PADDOCK, F.B.   1922   Foulbrood. Bull 97 8p illus
6   PADDOCK, F.B.   1926   Control of bee diseases and pests.
Bull 138 16p illus
     1929 Bull 154 23p
7   PADDOCK, F.B.   1929   Preparation of honey for sale. Bull 153
12p illus
     1936 rev 16p
8   PADDOCK, F.B.   1935   Control of American foulbrood. Circ 212
11p illus
     1941 rev 16p;   1950 Bull P105 [14p]
9   PADDOCK, F.B.   1939   Manual for beginners. In Report of the
State Apiarist for 1938 [62p] illus
     1943 Report for 1942 [78p];   1947   Report for 1947 [105p]
10   PAMMEL, L.H. & KING, C.M.   1930   Honey plants of Iowa.
Des Moines, Iowa Geological Survey Bull 7 1192p illus
11   PARK, O.W.   1928   Time factors in relation to the acquisition
of food by the honeybee. Res Bull 108 p. 184-225 illus bibliogr
     [See MD 6]
12   PARK, O.W.   1932   Studies on the changes in nectar concentration
produced by the honeybee, Apis mellifera. Part 1. Changes which occur
between the flower and the hive. Res Bull 151 p. 210-44 illus
13   PELLETT, F.C.   1913   Bee keeping in Iowa. Bull 11 20p illus
14   PELLETT, F.C.   1914   The beekeeper's library. State Bee
Inspector Bull 2 8p
     1915 rev Bull 4 13p
15   PELLETT, F.C.   1914   Brood diseases of bees. State Bee
Inspector Bull 3 12p illus From 2nd Annual Rep of State Bee Inspector
for 1913
     1932 PADDOCK, F.B. Bull 183 8p
     1939 rev Diseases and pests of bees. 16p [Excerpts from IOWA 6]
16   PELLETT, F.C.   1914   Wintering bees in Iowa. Bull 22 20p
illus
     1927 PADDOCK, F.B. Bull 141 8p no illus
     1936 rev 16p

KANSAS   Kansas State University, Manhattan, Kans. 66504

1   HUNTER, S.J.   1899   Alfalfa, grasshoppers, bees: Their
relationship. Lawrence, Univ Kans., Contribution from Entomol Lab 65
p. 65-152 illus
2   HUNTER, S.J.   1899   The honey-bee and its food-plants, with
special reference to alfalfa. Lawrence, Univ Kans., Dep Entomol Bull 7
85p illus
3   JOHNSON, J.A., MILLER, D. & WHITE, J.W., Jr.   1959   Honey in
your baking. Circ 281 23p illus recipes
4   MERRILL, J.H.   1918   Treatment of brood diseases of bees.
Topeka, Bd Agr, Circ 5 8p
     List of papers by Merrill publ in Kans. State Coll Bull 21(9): 22-24
     (1937)
5   MERRILL, J.H.   1918   Wintering bees. X-Form 65 2p
6   MERRILL, J.H.   1919   Making a start with bees. X-Form 106
2p

KANSAS

7   MERRILL, J.H.   1919   Spring management of bees. X-Form 95,
Circ 6 2p
8   MERRILL, J.H.   1924   Inspection and care of bees. Entomol
Comm Circ 13p illus
9   MERRILL, J.H. & HILL, F.   1922   Bees for the horticulturist:
Beekeeping in Kansas by J.H. Merrill p. 3-15; Comb honey by F. Hill
p. 15-23. Topeka, Kans Hort Soc
10   MILLER, D., WHITE, J.W., Jr. & JOHNSON, J.A.   1960   Honey
improves baked products. Bull 411 22p illus recipes
11   MOFFETT, J.O. & PARKER, R.L.   1953   Relation of weather
factors to nectar flow in honey production. Tech Bull 74 27p illus
11a   PARKER, R.L.   1932   Care of bees. Circ 10 p. 12-17
    List of papers by Parker publ in Kans. State Coll Bull 21(9): 30-31
    (1937)
12   PARKER, R.L.   1950   Major honey plant areas of Kansas.
Circ 81 2p
13   PARKER, R.L.   1953   Bee culture in Kansas. Bull 357 81p
illus
    1962 rev
14   TILTON, E.W. & PARKER, R.L.   1958   Grading Kansas and other
United States extracted honeys. Rep of Progress 25 24p mimeo

KENTUCKY   University of Kentucky, Lexington, Ky.   40506

1   BRAY, C.H. & EATON, W.G.   [1966?]   Honey bees and Kentucky
honey. Frankfort, Dep Markets 13p
2   DEP AGR, DEP COMMERCE & STATE BEEKEEPERS' ASS   1963   Honey
production in Kentucky, with recommended management practices and suggested
marketing and promotional tips for Kentucky beekeepers. Frankfort 28p
illus
3   GARMAN, H.   1917   Foul-brood of bees; its recognition and
treatment. Circ 17 [10p] illus
4   GARMAN, H.   1925   Beekeeping for beginners. Circ 35 34p
illus
5   KOEPPER, J.M.   [1955]   Kentucky 1955 honey production and
marketing survey. Frankfort, Dep Agr 29p illus
6   NISWONGER, H.R.   1919   Elements of beekeeping. Circ 69 22p
illus
    1922 rev 23p
7   PRICE, W.A.   1937   Beekeeping in Kentucky. Circ 288 35p
illus
    1941;   1949 rev 36p
8   SCHEIBNER, R.A.   1967   Beginning beekeeping for Kentuckians.
Misc Pub Ext Ser 361 28p

LOUISIANA   Louisiana State University, Baton Rouge, La.   70803

1   COCKERHAM, K.L. & OERTEL, E.   1952   Control of ant in
apiaries. 3p mimeo
2   OERTEL, E.   1940   A discussion of the natural history,
management and diseases of the honey bee. Baton Rouge, State Dep Agr
& Immigration 43p illus
    1947 2nd ed The beginner beekeeper in Louisiana. 52p
    1955 3rd ed 46p
3   STRONG, R.G.   1949   Beekeeping and insecticides. Ext Ser
10p mimeo

MAINE   University of Maine, Orono, Maine   04473

1   DIRKS, C.O.   1936   Beginning with bees in Maine. Bull 226
28p illus
    1946 Keeping bees in Maine. Bull 346 33p;   1952 rev 30p
2   GRIFFIN, O.B.   1918   Bee culture in Maine. Waterville,
Dep Agr Bull Vol 17(4) 58p illus

MARYLAND   University of Maryland, College Park, Md. 20742

1   ABRAMS, G.J.   1933   Honey bees for orchard pollination.
Bull 69 19p
2   ABRAMS, G.J.   1948   Migratory bee culture in Maryland.
Bull A48 21p
3   BURDETTE, R.F. & DEVAULT, S.H.   1939   The production and
marketing of honey in Maryland. Bull 427 30p illus
4   CALE, G.H.   [1917?]   Honey production, urgent necessity for
extra honey. Circ Letter 12 3p
5   CORY, E.N.   [1919?]   Status of beekeeping in Maryland.
Circ 38 4p
6   SYMONS, T.B. & McCRAY, A.H.   1911   Bee keeping in Maryland.
Bull 154 p. 227-69 illus bibliogr [See IOWA 11]

MASSACHUSETTS   University of Massachusetts, Amherst, Mass. 01003

1   BOURNE, A.I.   1927   The poisoning of honey bees by orchard
sprays. Bull 234 [12p] illus
2   DAVIS, A.M. & SHAW, F.R.   1938   Plants of value to Massachusetts beekeepers. Spec Circ 27 5p mimeo
    1950 rev Shaw, F.R. Honey and pollen plants of Massachusetts.
    1956 rev Nectar and pollen plants of Massachusetts with observations
    on the sugar concentration of certain nectars. 15p
3   FARRAR, C.L.   1929   Bees and apple pollination. Spec Circ 7
9p mimeo
4   GATES, B.N.   1908   Bee diseases in Massachusetts. Bull 124
12p [See F56 repr from Bull 75 (F11)]
5   GATES, B.N.   1909   Beekeeping in Massachusetts. Bull 129
32p illus [See F57 repr from Bull 75 (F11)]
6   GATES, B.N.   1910   Brood diseases of bees, their treatment and
the law for their suppression in Massachusetts. Boston, Bd Agr Apiary
Inspection Bull 1 12p
7   GATES, B.N.   1911   Warning to bee keepers and users of bees in
greenhouses. Boston, Bd Agr Apiary Inspection Bull 3 3p
8   GATES, B.N.   1912   Some of the essentials of beekeeping.
Boston, Bd Agr Apiary Inspection Bull 5 19p
    1915 2nd ed 22p;   1918 rev Everyday essentials of beekeeping.
    Bull 14 32p illus;   1923 rev Pub 121
9   GATES, B.N.   1914   Soft candy for bees. Boston, Bd Agr
Apiary Inspection Bull 7A 7p illus
10 GATES, B.N.   1916   Spraying versus beekeeping. Boston, Bd Agr
Apiary Inspection Bull 10A 22p illus
11 GATES B.N.   1917   Interpretation of the net weight regulations
for marking packages of honey. Boston, Bd Agr Apiary Inspection Bull 12
8p
12 HOWARD, S.M.   [1913]   Honeybees as pollinizers. A valuable
adjunct to the horticulturist. Boston, Bd Agr Apiary Inspection Bull 8
    1914 2nd ed 11p

MASSACHUSETTS

13   KELLOGG, C.R.   1933   Bees for the beginner in Massachusetts.
Leafl 148 32p illus
     1938 rev SHAW, F.R. 39p;   1942 rev 40p;   1947 rev 42p;
     1950 rev Beekeeping. 44p;   1957 rev 54p
14   PAIGE, J.B.   1907   The first principles of bee keeping.
Boston, Bd Agr Nature Leafl 34 10p illus
15   ROOT, E.R.   1918   The importance of honey production. Boston,
Bd Agr Circ 81 10p
16   SHAW, F.R.   195?-1958   [Circulars and Leaflets]
     nd Observation beehives. Construction - Use. Spec Circ 191 1p
     nd Bees - How to remove them from buildings. Circ 204 2p
     1955 rev Leafl 292 4p
     1954 Spring management of honey bees. Circ 241 3p
     1958 rev 4p
     1955 Bees for beginners. Spec Circ 252 2p
17   SHAW, F.R.   1965   Honeybees, their behavior, biology and value.
Pub 432 16p illus

MICHIGAN   Michigan State University, East Lansing, Mich. 48824

1   ANON   nd   Recipes to match your sugar ration. 8p
2   COOK, A.J.   1890   Foul brood. Bull 61 8p
3   COOK, A.J.   1890   Planting for honey. Bull 65 7p illus
4   COOK, A.J.   1893   Honey analyses. Bull 96 16p
5   FABIAN, F.W.   1926   Honey vinegar. Circ Bull 85 13p illus
     1935 Bull 149 14p
6   FABIAN, F.W. & QUINET, R.I.   1928   A study of the cause of
honey fermentation. Tech Bull 92 41p illus
7   HERSHEY, R.   1935   Hints for using honey. Bull 150 16p
illus
     1936 rev
8   HOOVER, R.H.   1940   Honey flavor harmonies. Bull 213 36p
illus recipes
9   JORGENSEN, C. & MARKHAM, F.   1946   Weather factors influencing
honey production. Spec Bull 340 22p illus
10   KELTY, R.H.   1921   Diseases of bees in Michigan. Spec Bull 107
15p illus
11   KELTY, R.H.   1922   Packing case for outdoor wintering.
Press Bull 66 2p
12   KELTY, R.H.   1922   Queen rearing for home use. Press Bull 67
4p
13   KELTY, R.H.   1922   Tar-paper packing-case. Press Bull 68 1p
     1929 ...for wintering bees. Bull 77 4p illus
14   KELTY, R.H.   1923   Beekeeping. Outline of course to be used
in Michigan high schools. 26p
15   KELTY, R.H.   1924   Seasonal management for commercial apiaries.
Spec Bull 135 58p illus suppl
     1929 rev 60p
     1941 Bull 228 92p Re-issued with Beekeeper's guide to seasonal
     management. Folder F-2 2p
16   KELTY, R.H.   1929   Renting or keeping bees for use in the
orchard. Bull 56 11p illus
17   KELTY, R.H.   1948   Observations on the performance of package
bee colonies in Michigan. Spec Bull 344 37p illus
18   KELTY, R.H. & HARRISON, C.M.   1950   The importance of the
honeybee to Michigan agriculture. Folder F-152 7 col

MICHIGAN

19   KINDIG, B.F.   1921-1926   Tips and pointers on successful
beekeeping. Lansing, Dep Agr [See MICH 28]
    1921 American foul brood. Bull 2 11p
    1922 Nectar producing resources of Michigan. Bull 4 21p
    1926 rep
    1922 European foul brood. Bull 5 4p
20   MARTIN, E.C.   1952   Basic beekeeping information. Folder F-165
8p
    1958 rev
21   MARTIN, E.C.   1968   Basic beekeeping. Bull E625 12p
22   MILLEN, F.E.   1915   Transferring bees. Spec Bull 76 16p
illus
    1923 rev KELTY, R.H. 12p
23   MOTTS, G.N.   1943   Marketing Michigan honey. Spec Bull 321
50p illus
    1961 Marketing handbook for Michigan honey. Bull 433 36p
24   PETTIT, R.H.   1912   Foul brood. Spec Bull 58 12p illus
    1914 MILLEN, F.E. Spec Bull 64 Suppl to Bull 58 8p
    1919 KINDIG, B.F. Circ 39 Suppl to Bull 58 & 64 3p
25   SHAFER, G.D.   1917   A study of the factors which govern
mating in the honey bee. Tech Bull 34 19p 3 plates
26   SMITH, C.D. & RANKIN, J.M.   1901   Foul brood. Spec Bull 14
11p
27   TAYLOR, R.L.   1894   A year with bees. Spec Bull 1 [26p]
28   ULMAN, P.T.   1922   Tips and pointers on successful beekeeping.
Lansing, Dep Agr [See MICH 19]
    1922 Requeening the apiary. Bull 3 10p illus

MINNESOTA   University of Minnesota, Minneapolis, Minn. 55455

1   CHILDS, A.M. & KOLSHORN, A.   1928   Honey; how to use it.
Spec Bull 122 8p illus
    1929;   1935 rev Childs, A.M. & NILES, K.B. 23p
2   HAYDAK, M.H.   1945   Pollen substitutes. Folder 130 6 col
illus
    1953 rev
3   HAYDAK, M.H. & TANQUARY, M.C.   1943   Pollen and pollen
substitutes in the nutrition of the honeybee. Tech Bull 160 23p
4   JAGER, F.   1919   Management of bees. Care of bees in spring.
Spec Bull 38 4p
5   JAGER, F.   1923   Habits and activities of bees. Spec Bull 73
20p illus
    1928 rev 24p
6   JAGER, F.   1923   Requeening the apiary. Circ 10 2p
7   JAGER, F.   1926   The bee industry. Radio Ext Courses, Radio
Ser No. 4
8   JONES, G.D. & MARTIN, T.T.   1939   The 4-H beekeepers' club.
Entomol II. 4H Club Circ 61 20p illus
9   MATTHEWS, G.C.   1920   Queen rearing. Spec Bull 49 8p illus
10   ROWE, I.B. & TANQUARY, M.C.   1942   Honey for everyday use.
Bull 239 8p
11   TANQUARY, M.C.   1939   Beekeeping in Minnesota. Bull 204
32p illus
    1948 rev HAYDAK, M.H. 37p;   1955 rev 39p;   1962 rev 40p

MISSISSIPPI   State Plant Board, State College, Miss.   39762

1   LYLE, C.   1941   Beekeeping in Mississippi. Circ 1 (n.s.)   12p
   1958 rev TATE, H. & WILSON, C.A.   20p

MISSOURI   University of Missouri, Columbia, Mo.   65202

1   BIRKETT, T.E.   1942   Bees to fill the sugar bowl. Circ 462
4p
   1943
2   CARL, F.L. & KNIGHT, L.O.   1942   Using sugar, honey, sorghum
and other sirups. Circ 467   12p
3   DARBY, M.E.   nd   Bees and horticulture. Bull 42 6p
4   HASEMAN, L.   1931   Beekeeping in Missouri. Bull 305 52p
illus
   1939;   1947 rev
5   HASEMAN, L.   1949   Sulfathiazole control of American foulbrood.
Circ 341 8p illus
6   HASEMAN, L. & CHILDERS, L.F.   1944   Control of American foulbrood with sulfa drugs. Bull 482 16p
   1946
7   TYLER, E.E. & HASEMAN, L.   1915   Farm bee-keeping. Bull 138
40p illus

MONTANA   Montana State University, Bozeman, Mont.   59715

1   BENTON, R.   1907   Practical beekeeping. Bull 67 75p illus
2   SIPPEL, O.A.   1923   Bee diseases in Montana. Circ 120 14p
illus

NEBRASKA   University of Nebraska, Lincoln, Nebr.   68508

1   BESSEY, C.E.   1895   A preliminary list of the honey-producing
plants of Nebraska. Bull 40 Vol 7, Article 4 p. 141-51
2   BOARD AGR   1912   Fruit growing and bee culture in Nebraska.
Bull 24 16p
3   HELM, R.W.   1951   Removing bees from a building. EC 1566   4p
4   HIXSON, E.   1955   Insects that affect seed production in
Nebraska. Bull 433 20p illus
5   PETERS, M. & ATWOOD, F.J.   1933   Honey thruout the year.
Circ 911 16p illus recipes

NEW HAMPSHIRE   University of New Hampshire, Durham, N.H.   03824

1   HEPLER, J.R.   1934   Beekeeping in New Hampshire. Bull 46
16p illus
   1948 Bull 86 24p
2   WHITCOMB, W.   1924   American foulbrood. Press Bull 137 1p
   1924 European foulbrood. Press Bull 138 1p
3   WOLFF, W.H.   1921   Beekeeping for New Hampshire. Bull 15
16p illus

NEW JERSEY   Rutgers University, New Brunswick, N.J.   18903

1   CARR, E.G.   1915   A manual of bee husbandry. [Trenton],
Board Agr 72p illus
   1922 Dep Agr Circ 53 91p;   1928 Rutgers Univ Bull 463 87p;
   1934 Carr, E.G. & FILMER, R.S. Circ 317 88p;
   1941 Filmer, R.S. Circ 404 79p
2   CARR, E.G.   1917   Bee disease control. Trenton, Dep Agr
Circ 3 30p illus
3   CARR, E.G.   [1918]   Profitable beekeeping. (in 11 parts)
[Trenton], Dep Agr 29p

NEW JERSEY

4   CARR, E.G.   1925   Directions for the treatment of European foulbrood. Trenton, Dep Agr Circ 82 4p; Directions for the treatment of American foulbrood. Circ 83 4p

5   CARR, E.G.   1931   Bee disease control work in New Jersey. Trenton, Dep Agr Circ 197 16p

6   CARR, E.G.   1932   Transferring bees to movable-comb hives. Circ 217 6p

7   CARR, E.G.   nd   The treatment of American foul brood. Trenton, Dep Agr Circ 225
     1941 HOLCOMBE, P.L. Circ 266 9p;   1943 rev 10p illus;
     1947 Circ 367 9p;   1952 rev Circ 387;
     1958 MATTHENIUS, J.C. American foul brood and its treatment. Circ 408 11p

8   DOERMANN, M.C.   1932   Honey - Its use in cookery. Bull 99 8p recipes
     1933 Bull 103

9   FED WRITERS' PROJECT   1938-39 Ser   Stories of New Jersey. Bees and their work. For use in public schools. Bull 6 4p illus

10   FENTON, J.M.   1929   The cost of producing honey in New Jersey and other economic data on beekeeping. Trenton, Dep Agr Circ 159 44p

11   FILMER, R.S.   1942   Preparing bees for winter. Circ 424 4p; Package bees. Circ 426 4p

12   FILMER, R.S.   1944   A study of the wintering problem in New Jersey. Exp Sta Notes 6p mimeo

13   FILMER, R.S. & DOEHLERT, C.A.   1952   Use of honeybees in cranberry bogs. Bull 764 4p
     1959 Circ 588 4p

14   FILMER, R.S. & SWIFT, F.C.   1963   Blueberry pollination - lesson 1. Leafl 359 4 col

15   HUTSON, R.   1926   Relation of the honeybee to fruit pollination in New Jersey. Bull 434 32p

16   HUTSON, R.   1929   Chemicals in the apiary with notes on their use. Circ 211 8p

17   HUTSON, R.   [1930]   The races of honeybees and their characteristics. 4p

18   McEVOY, W.   1895   Foul brood. Its cause and cure. Trenton, Board Agr 8p

19   MORRISON, W.C.   1957   Woody honey plants for roadside planting in New Jersey. Trenton, Dep Agr Circ 403 23p bibliogr

20   PITT, D.T. & CARR, E.G.   1935   The beekeeping industry in New Jersey. Trenton, Dep Agr Circ 247 21p illus
     1937 Pitt, D.T. Circ 279 104p
     1946 The New Jersey beekeeping industry in 1944. Circ 362 54p

NEW YORK   Cornell University, Ithaca, N.Y. 14850

1   ALEXANDER, W.P.   1918   Beginnings in beekeeping. Lesson 138 Cornell Reading Course for the Farm p. 75-96 illus

2   BUR COOPERATIVE ASS   [1921]   Explanation of the plan of organization of the Empire State Honey Marketing Cooperative Association, Inc. Albany, Dep Farms & Markets 15p

3   COGGSHALL, W.L.   [1950-51]   [Ext folders]
     Bees make more than honey. 5 col illus
     Some pollination suggestions. 2p illus
     Spring management of bees. 6 col illus
     There are lots of ways to work with bees. 1 sheet illus

NEW YORK

4   COLL AGR   [1953?]   Home study course in beekeeping. Includes
Study manual for beekeeping. 53p 13 lessons & 3 practical exercises.
5   DEP AGR   1907   Disease among bees and treatment. Albany 16p
illus
6   DEP FARMS & MARKETS   [1916?]   Diseases of the honeybee and
treatment. Albany Circ 97 11p 2 plates
    1925 ATWOOD, G.G. Circ 282 10p
    1927 VAN BUREN, B.D. Circ 325 16p illus
7   DOBROVSKY, T.M.   1951   Postembryonic changes in the digestive
tract of the worker honeybee (Apis mellifera,L.). Memoir 301 45p 12 plates
8   DYCE, E.J.   1931   Fermentation and crystallization of honey.
Bull 528 76p
9   DYCE, E.J.   1951   Beekeeping. General information. Bull 833
18p
    1955 rev 15p;   1958 rev 16p;   1962 rev 15p
10   DYCE, E.J.   1960   Removing colonies of honey bees from
buildings and other undesirable places. Dep Entomol 4p
11   DYCE, E.J. & MORSE, R.A.   1960   Wintering honeybees in New
York state. Bull 1054 16p illus
    1962 rev
12   GOULD, A.C.   1929   Diseases of the honey bee and their control.
Albany, Dep Agr & Markets Circ 381 18p illus
    1931 Circ 429;   1938 Circ 553 20p
13   KING, E.R.   1917   How to increase the honey supply. Bull 16
p. 537-43
14   MILLER, H.W. & WOODS, B.   1957   Honey. Food Mktg Handbook,
Sugars & Sweets 2 11p
15   MOORE, V.A. & WHITE, G.F.   1903   A report on the investigation
of an infectious bee disease. Albany, Dep Agr 10p 2 plates
    ["N.Y. bee disease" or "black brood"]
16   MORSE, R.A.   1954   Bibliography of honey wine. 5p mimeo
17   MORSE, R.A.   1960   The scope of the beekeeping industry in
New York state. 5p
18   NORTON, G.C.   1922   Bee-keeping. Farmingdale, N.Y. State Inst
Applied Agr Vol 7(1) [32p] illus
19   PARKER, R.L.   1926   The collection and utilization of pollen
by the honeybee. Memoir 98 55p illus bibliogr
20   PHILLIPS, E.F.   1929   Variation and correlation in the
appendages of the honeybee. Memoir 121 52p illus
21   PHILLIPS, E.F.   1930   Honeybees for the orchard. Bull 190
24p illus
    Also publ as Bees and fruit. Medina, [A.I. Root Co] 16p illus
22   RASMUSSEN, M.P.   1932   Some facts concerning the production
and marketing of honey. Bull 221 110p
23   REA, G.H.   1940   Handling package bees. Bull 433 8p illus;
    Honeybees for pollination. Bull 434 4p; Spring management of the
    apiary. Bull 436 8p
24   SINGH, S.   1950   Behavior studies of honeybees in gathering
nectar and pollen. Memoir 288 57p illus
25   SMITH, M.V.   1959   Queen differentiation and the biological
testing of royal jelly. Memoir 356 56p illus
26   WILLSON, R.B. & FAIRBANKS, F.L.   1925   Honey and water as an
antifreezing solution for automobiles. Mimeo Bull R2 1p
27   WOODROW, A.W.   1933   The comparative value of different
colonies of bees for fruit pollination. Memoir 147 29p

NEW YORK

28   WRIGHT, W.D.   1913   The honey bee. Albany, Dep Agr Bull 49
p. 1383-535 illus

NORTH CAROLINA   North Carolina State University, Raleigh, N.C. 27607

1   CARR, E.G.   1917   Beekeeping practice in North Carolina.
Circ 52 8p
2   SHERMAN, F.   1908   Bee-keeping in North Carolina. A study of
some statistics on the industry, with suggestions and conclusions.
Bull Dep Agr Vol 29(1) 27p illus
3   STATE MUS   nd   Important honey plants of North Carolina.
Circ 46-5 1p mimeo
4   STEPHEN, W.A.   nd   4-H honey production record book. 19p
[1953?] rev 4-H beekeeping record book. 8p
5   STEPHEN, W.A.   1949   Bee lines. Circ 334 30p illus
1955 rev 31p;   1962 rev;   1964
6   STEPHEN, W.A.   1963   Beekeeping. Ent. 322
Correspondence course
7   STEVENS, R.O.   nd   Practical beekeeping. Circ 274 24p illus

NORTH DAKOTA   North Dakato State University, Fargo, N.Dak. 58102

1   LEEBY, C.   1931   Honey cookery. Circ 108 12p illus
2   MUNRO, J.A.   [1927]   Bees in North Dakota. Bismark, Dep Agr
& Labor 8p
1929 ..., the land of milk and honey. 23p illus
3   MUNRO, J.A.   1932   Bee inspection in North Dakota. Bismark,
Dep Agr & Labor 11p illus
4   MUNRO, J.A.   1932   Bees and their care. Circ 112 16p illus
1939 rev
5   POST, R.L. & COLBERG, W.J.   1959   Beginning beekeeping.
Circ A-310 12p illus

OHIO   Ohio State University, Columbus, Ohio 43210

1   ARGO, V.N.   1930   Handling package bees. Bull 98 12p illus
2   BEARD, D.F. & DUNHAM, W.E.   1944   Honeybees increase clover
seed production 15 times. Bull 253 4p illus
1945 rev;   1948 rev Beard, D.F., Dunham, W.E. & REESE, C.A.;
1950
3   DUNHAM, W.E.   [1933]   Beekeeping. Correspondence Courses
in Agr. Course XVI. [80p] mimeo illus 8 lessons
[1953 rev];   nd HINE, J.S. 6 lessons
1966 rev STEPHEN, W.A. Ext Ser No. L49 12 lessons
4   DUNHAM, W.E.   1935   Package bees for honey production.
Bull 159 12p illus
1938 rev 16p;   1942 rev;   1946 rev;
1951 rev REESE, C.A. ...and pollination. 12p
5   DUNHAM, W.E.   1940   Modified two-queen system for honey
production. Bull 281 4p illus mimeo
1941 6p;   1947 16p
6   DUNHAM, W.E.   1944   Bees; maintenance of colonies, control
of colony population for honey production and pollination. Bull 254
32p illus
1947 Bees for honey production and pollination
1952 rev REESE, C.A. 16p
7   DUNHAM, W.E.   1947   Honey and honey cookery. Bull 282 31p
illus

OHIO
8   ROOT, E.R.   1913   Bees. Bull Vol 9(3) 16p illus
9   SCULLEN, H.A.   1923-24   Beekeeping in Ohio. Bull 32p illus
   Adapted from "Beekeeping in Oregon". Bull 360 [See OREG 9]
10   STEPHEN, W.A.   [1964]   Ohio bee lines. Bull 450 28p illus
11   WEBSTER, F.M.   1896   Some destructive insects; spraying with
arsenites vs. bees; carnivorous habits of _Limax_ _campestris_. Bull 68
2nd Ser 58p [bees p. 48-53]

OKLAHOMA   Oklahoma State University, Stillwater, Okla. 74075
1   ANON   1908   Bee culture in Oklahoma. Press Bull 155 1p
2   BIEBERDORF, G.A.   1954   Honeybees as vetch pollinating agents.
Mimeo Circ M-250 3p
3   OSBORN, H.   1909   Some things to learn about bees. Bull Vol
4(9) [4p]
4   RUDE, C.S.   [1927]   Principles of beekeeping for beginners.
Circ 238 Gen Ser 77 16p illus
    1932 rev STILES, C.F. 24p;   1936 rev 29p;   1943 rev 30p;
    [1948] 29p;   [1953?] SHAW, F.R. & WHITEHEAD, S.B. Beekeeping for
    beginners. 24p
5   SANBORN, C.E.   1914   Information relative to beekeeping.
Circ 28 8p
6   SANBORN, C.E.   1916   How every farmer can keep a few bees and
have honey at very little cost. Circ 34 [4p] illus
7   SANBORN, C.E.   1917   Methods by which, with a few colonies of
bees, honey may be had throughout the year. Circ 52 4p
8   SANBORN, C.E.   1918   Beekeeping. Circ 74 [2p]
9   SANBORN, C.E.   1919   Beekeeping in Oklahoma. Circ 106 26p
illus
10   SANBORN, C.E.   1920   Winter treatment for honey bees. Circ
of Inform 48 [4p] illus
11   STILES, C.F.   [1943?]   Beekeeping. OP-41 5 col folder

OREGON   Oregon State University, Corvallis, Oreg. 97331
1   BURRIER, A.S., TODD, F.E., SCULLEN, H.A. & GORTON, W.W.   1939
Costs and practices in producing honey in Oregon. Bull 362 38p
2   DEP FOODS & NUTRITION   1934   Uses of honey. Bull 472 16p
recipes
3   LOVETT, A.L.   1922   The brood diseases of bees. Circ 82 5p
4   PRENTISS, S.W.   1920   Honey, nature's oldest sweet. Bull 322
3p
5   SCULLEN, H.A.   1922   A simple method of queen rearing for the
small beekeeper. Circ 196 2p
6   SCULLEN, H.A.   1956   Bees... for legume seed production.
Circ 554 15p illus
7   SCULLEN, H.A. & VANSELL, G.H.   1942   Nectar and pollen plants
of Oregon. Bull 412 63p illus
8   STEPHEN, W.P.   1957   Diseases of bees... How to recognize
and control. Circ 629 8p illus
    1961 rev
9   WILSON, H.F.   1914   Beekeeping for the Oregon farmer. Bull 168
Ser 2(25) 31p illus
    1919 rev LOVETT, A.L. Bull 282 24p;
    1923 rev SCULLEN, H.A. Beekeeping in Oregon. Bull 360 32p;
    1927 rev Bull 401;   1930 rev Bull 430;   1933 rev Bull 462;
    1938 rev Bull 513;   1943 rev Bull 622 36p

PENNSYLVANIA   Pennsylvania State University, University Park, Pa. 16802

1   ANDERSON, E.J.   1931   Beekeeping in Pennsylvania. Circ 141
36p illus
     1936 rev 42p;   1942 2nd rev 54p;   1946 3rd rev 62p;
     1950 Anderson, E.J. & CLARKE, W.W., Jr. Circ 357 54p
2   ANDERSON, E.J.   1935   Diseases and enemies of the honeybee.
Circ 156 16p illus
     1941 rev 20p
3   ANDERSON, E.J.   1937   Extracting equipment for the honey
house. Circ 188 23p illus
     1943 rev 25p
4   ANDERSON, E.J.   1944   A word about package bees. Exp Sta
Paper 1225, J Ser 3p mimeo
5   ANDERSON, E.J.   1954   Honey candies and honey spreads. 16p
mimeo recipes
     1958 Honey candies. Progress Rep 186 4p;   1960 rev
6   ANDERSON, E.J.   1960 .   An improved solar beeswax extractor.
Exp Sta Progress Rep 225 5p illus
7   CLARKE, W.W., Jr.   [1957]   Diseases of bees and their control.
Circ 470 13p illus
8   CLARKE, W.W., Jr. & ANDERSON, E.J.   [1957]   Pennsylvania
beekeeping. Circ 472 57p illus
     1965 53p;   [1969] Clarke, W.W., Jr. Circ 544 63p
9   DEP AGR   1926   Publications on beekeeping. Harrisburg Circ 5
2p mimeo
10   KIRK, H.B.   1931   Apiary inspection in Pennsylvania. Harrisburg, Dep Agr Gen Bull 499 15p illus
11   MILLER, C.C.   1901   Bee culture. Harrisburg, Dep Agr Bull 77
103p illus
12   SURFACE, H.A.   1912   Bee-keeping. Harrisburg, Dep Agr
Zoological Bull 2(3) p. 80-151 illus
13   SURFACE, H.A. & REA, G.H.   1916   Bee diseases in Pennsylvania.
Harrisburg, Dep Agr Zoological Bull 6(4) [23p]

RHODE ISLAND   University of Rhode Island, Kingston, R.I. 02881

1   CUSHMAN, S.   1889   Beekeeping. Bull 4 p. 71-97
2   CUSHMAN, S.   1890   Report of the apiarist. Bull 7 p. 60-63
3   CUSHMAN, S.   1890   Experiments in apiculture. Foul brood,
its cause, prevention and cure. Bull 9 p. 101-21 illus
4   MILLER, A.C.   1911   Bee keeping in Rhode Island. Providence,
Board Agr 11p
5   MILLER, A.C.   1911   How to keep bees. Providence, Board Agr
39p illus bibliogr
     1918 rev 50p
6   PIKE, H.A.   1947   Beekeeping in Rhode Island. Providence,
Dep Agr 80p illus

SOUTH CAROLINA   Clemson University, Clemson, S.C. 29631

1   DEP AGR EDUC   1939   Beekeeping and honey production. Bull
Vol. 15(11/12) p. 163-92
2   PREVOST, E.S.   1936   The care of bees. Circ 153 15p illus
     1943
3   PURSER, W.H. & SPARKS, L.M.   1962   Beekeeping in South
Carolina. Bull 122 29p illus

<u>SOUTH DAKOTA</u>   South Dakota State University, Brookings, S. Dak. 57007

1   WALSTROM, R.J. & HOUTSBARGER, W.M.   1959   Beekeeping in South
Dakota. Circ 565 16p illus

<u>TENNESSEE</u>   Department of Agriculture, Nashville, Tenn. 37219

1   BARTHOLOMEW, C.E.   1917   Wintering bees in Tennessee. Coll
Agr Pub 53 8p
2   BENTLY, G.M.   1913   Beekeeping in Tennessee. Bull 9 64p
illus
3   BUCHANAN, J.M., DEMUTH, G.S., HEATHERLY, J.M. & ALLEN, G.   1922
Beekeeping in Tennessee. Tenn Agr Vol. 5(2) 26p illus
4   LITTLE, L.H. & WALLACE, L.D.   1952   A bee book for bee-ginners.
    1953 2nd ed;   1955 3rd ed;   1958 4th ed;   1961 5th ed;
    1967 6th ed rev

<u>TEXAS</u>   Texas Agricultural & Mechanical University, College Station,
         Tex. 77843

1   FRAPS, G.S.   1921   The chemical composition of Texas honey and
pecans. Bull 272 9p
2   GREGG, P. & LITTLE, V.A.   1947   Beekeeping for beginners.
B-153 20p illus
    1962 rev BURGIN, C.J., GARNER, C.F.,& RIDGWAY, R.L.   An introduction
    to beekeeping.
3   HERRICK, G.W. & SCHOLL, E.E.   1908   The foul brood of bees
and the foul brood law. Bull 116 10p
4   MALLY, F.W. & NEWELL, W.   1902   Report on the A. & M. College
apiary; together with practical suggestions in modern methods of bee
keeping as applied to Texas conditions. 53p illus
    1911 rev Newell, W. Practical information for beginners in beekeeping.
    Bull 142 46p
    Superseded by Bull 255 [TEX 6]
5   NEWELL, W., PADDOCK, F.B. & DEAN, W.H.   1913   Investigations
pertaining to Texas beekeeping. Bull 158 43p illus
    Experiments in artificial division and swarm-control by Newell;
    The life history and control of the bee-moth or wax worm by Paddock;
    1918 rev The beemoth or waxworm. Bull 231 38p
    A statistical study of Texas beekeeping by Dean
6   PARKS, H.B.   1919   Beekeeping for beginners. Bull 255 25p
illus
    1924 28p;   1935
    Superseded Bull 142 [TEX 4]
7   PARKS, H.B.   1927   Legumes for Texas beekeepers. Circ 46
11p illus
8   PARKS, H.B.   1927   The work of the state apicultural research
laboratory 1919-1926. Bull 361 16p illus
9   PARKS, H.B. & ALEX, A.H.   1925   Suggestions on queen rearing.
Circ 35 19p illus
10   PINCKNEY, J.M. & MILLER, A.H.   1931   Activities of the honeybees. Unit of instruction in health educ. Austin, Univ Tex. 80p
11   REPPERT, R.R.   1931   Demonstration in bee keeping. Circ 86
8p illus
    1940 Elementary beekeeping.
12   SANBORN, C.E. & SCHOLL, E.E.   1908   Texas honey plants.
Bull 102 31p

-83-

TEXAS

13   SCHOLL, L.H.   1912   Texas beekeeping. Austin, Dep Agr Bull 24
142p illus bibliogr
     1912 (March-April) 2nd ed
14   THOMAS, F.L. & RUDE, C.S.   1924   I. Foulbrood control and
diseases of bees. II. Foulbrood law and revised regulations. Circ 36 24p
illus

UTAH   Utah State University, Logan, Utah 84321

1   BOHART, G.E. & KNOWLTON, G.F.   1951   Honey bees for higher
yields of alfalfa seed in Utah. Bull 238 5 col folder illus
     1952 Circ 154
2   KNOWLTON, G.F.   1954   Growing alfalfa seed without destroying
pollinators. Circ 155 2p illus
3   KNOWLTON, G.F.   1954   Safeguard bees during orchard spraying.
Circ 212 2p illus
4   KNOWLTON, G.F. & NYE, W.P.   1952   Raising bees. Circ 157
2p illus
     1954
5   KNOWLTON, G.F. & SORENSON, C.J.   1947   Alfalfa seed production
requires pollination by bees. Bull 150 4p illus
6   KNOWLTON, G.F., STURTEVANT, A.P. & SORENSON, C.J.   1950   Adult
honey bee losses in Utah as related to arsenic poisoning. Bull 340 30p
illus
7   LEVIN, M.D. & NYE, W.P.   1951   Feeding pollen supplement and
pollen substitute to honeybees. Bull 237 5 col folder illus
8   VANSELL, G.H.   1949   Pollen and nectar plants of Utah.
Circ 124 28p illus

VERMONT   University of Vermont, Burlington, Vt. 05401

1   BEECHER, H.W.   1910   The honey bee in Vermont. Montpelier,
Dep Agr Bull 6 29p
2   CRANE, J.E.   1914   Bee diseases and their treatment. Montpelier, Dep Agr Bull 18
     With H.L. Bailey Formulae for tree spraying when in foliage.
3   ROCK, J.P.   nd   Bee disease in Vermont. Inspection Serv 3p
4   TOMPKINS, E.H.   1961   Honey production and marketing in
Vermont. Bull 618 40p illus

VIRGINIA   Virginia Polytechnic Institute, Blacksburg, Va. 24061

1   GRAYSON, J.M. & ROWELL, J.O.   1949   Beekeeping in Virginia.
Bull 178 29p illus
     1950 rev;   1954 rev 31p
2   KOINER, G.W.   1917   Bee culture in Virginia. Richmond, Board
Agr Bull 115 9p

WASHINGTON   Washington State University, Pullman, Wash. 99163

1   CENTRAL WASH. BEEKEEPERS ASS   nd   M-m-m honey. Yakima, Dep
Agr 13p mimeo recipes
2 CREWS, A.   1935   How to conduct the 4-H bee club. 8p
3   GOWAN, F. & SLOCUM, B.A.   1924   Wintering bees in western
Washington. Ext Serv 122 11p illus
4   JOHANSEN, C.   1959   Bee poisoning; A hazard of applying
agricultural chemicals. Circ 356 5p
     1960 rev;   1961 rev 7p;   1963 rev 9p
5   JOHANSEN, C.   1966   Beekeeping. Ext Pub PNW 79 27p

WASHINGTON

6   JOHANSEN, C.   1969   The bee poisoning hazard from pesticides.
Bull 709 8p 6p tables
7   JOHANSEN, C.A., LEVIN, M.D., EVES, J.D., FOSSYTH, W.R., BUSDICKER,
H.B., JACKSON, D.S. & BUTLER, L.I.   1965   Bee poisoning hazard of
undiluted malathion applied to alfalfa in bloom. Circ 455 10p
8   SLOCUM, B.A.   1921   Control of European and American foul
brood and wintering of bees. Bull 75 8p
9   SLOCUM, B.A.   1922   Marketing honey. Bull 88 6p
10   SLOCUM, B.A.   1923   Honey and its uses. Bull 94 19p recipes
11   SLOCUM, B.A.   1925   Disinfectants for European and American
foulbrood. Bull 128 10p illus
12   SLOCUM, B.A.   1926   Transferring bees to movable frame hives.
Bull 135 3p illus
   1928
13   SLOCUM, B.A.   1927   European foulbrood. Bull 144 6p illus
14   SLOCUM, B.A.   1927   Swarm control. Bull 145 10p illus
15   SLOCUM, B.A.   1927   American foulbrood. Bull 147 7p illus
   1934 rev Bull 195
16   WEBSTER, R.L.   1940   Control of bee diseases. Bull 256 8p
illus
17   WEBSTER, R.L.   1942   Beekeeping in Washington. Bull 289
26p illus
   1945 rev 37p;   1947 rev 44p;   1957 rev Webster, R.L. &
   JOHANSEN, C. 24p;   1961 rev Johansen, C.
18   WEBSTER, R.L., TELFORD, H.S. & MENKE, H.F.   1949   Bees and
pollination problems. Circ 75 8p

WEST VIRGINIA   West Virginia Department of Agriculture, Charleston,
                 W. Va. 25305

1   ANON   nd   Honey in the rock cook book. 40p
2   REESE, C.A.   1917   Beekeeping for West Virginia. Bull 33
52p illus

WISCONSIN   University of Wisconsin, Madison, Wis. 53706

1   DEP ENTOMOL   1957   How to destroy bees in the home. Circ 547
4p illus
2   FRANCE, N.E.   1902   Wisconsin bee-keeping. Platteville,
State Inspector Apiaries Bull 2 68p illus
3   FRANCE, N.E. & FRANCE. L.V.   1916   Beekeeping in Wisconsin.
Bull 264 28p illus
   N.E. France, Assistant in Entomol, Wis. Expt Sta 1914-15; L.V. France,
   Instructor in Beekeeping, Univ. of Minn.
4   HAMBLETON, J.I.   1924   Beekeeping in Wisconsin. Circ 174
24p illus
5   JONES, B.B.   1920   Standards for grades of honey. State Dep
Agr Bull Vol 1(3) 4p
   1922 Jones, B.B. & ADAMS, C.D. Grading and marketing Wisconsin
   honey. Bull Vol 3(4) 19p illus
   1927 POMERENING, A.W. & Adams, C.D. Standard grades of Wisconsin
   honey. Bull Vol 8(2) 4p
   1931 Wisconsin standards for the grading and packing of Wisconsin
   honey. Bull 127 16p
6   MAKAR, S.   1964   New concept for pollen trapping. Bull 568
7   RAHMLOW, H.J.   1968   Beekeeping in Wisconsin. Research has
changed our honey producing practices. Circ 659 20p

WISCONSIN

8   WEBER, P.D.   1956   Wisconsin honey production and marketing.
Dep Agr Spec Bull 61 40p illus
9   WHITCOMB, W. & WILSON, H.F.   1929   Mechanics of digestion of
pollen by the adult honey bee and the relation of undigested parts to
dysentery of bees. Res Bull 92 27p illus
10   WILSON, H.F.   1916   Better queens produce better bees. Stencil
Bull 11 3p
    1920
11   WILSON, H.F.   1921   How to control American foulbrood.
Bull 333 21p illus
12   WILSON, H.F.   1922   Winter care of bees in Wisconsin.
Bull 338 26p
13   WILSON, H.F.   1933   Beekeeping methods for Wisconsin.
Circ 258 31p illus
14   WILSON, H.F. & MILUM, V.G.   1927   Winter protection for the
honey bee colony. Res Bull 75 47p illus

WYOMING   University of Wyoming, Laramie, Wyo. 82071

1   CORKINS, C.L.   1928   The use of calcium cyanide in the apiary.
Bull 158 p. 109-16
2   CORKINS, C.L.   1930   The metabolism of the honeybee colony
during winter. Bull 175 52p illus
3   CORKINS, C.L.   1932   Drifting of honeybees. Bull 190 22p
illus
4   CORKINS, C.L. & GILBERT, C.H.   1932   A comparative test of the
Caucasian with the Italian race of honeybees. Bull 186 24p illus
5   CORKINS, C.L. & GILBERT, C.H.   1932   The metabolism of honeybees in winter cluster. Bull 187 30p illus
6   GILBERT, C.H.   1928   Surface tension of disinfecting solutions
for American foulbrood. Bull 159 p. 119-31 illus
7   GILBERT, C.H.   1929   The sterilization of brood combs infected
with American foulbrood. Bull 166 p. 151-69
8   GILBERT, C.H.   1939   The cellar wintering of bees. Bull 234
18p illus
9   GILBERT, C.H.   1940   Wintering bees in Wyoming. Bull 238
15p illus
10   GILBERT, C.H.   1940   The two-queen hive and commercial honey
production. Bull 239 15p illus
11   ROKAHL, M. & CORKINS, C.L.   1924   Wyoming honey for Wyoming
people. Circ 15 4p recipes
    1925 SHERMAN, L. Circ 17 [3p]

BEEKEEPING ASSOCIATIONS

International. The International Federation of Beekeeper's Associations (Apimondia) sponsors the International Apicultural Congresses (1st in Brussels in 1897, 23rd in Moscow in 1971) and publishes complete proceedings of the meetings. Five standing commissions carry on the work of the organization between Congresses and a press was established in 1966 to publish the journal _Apiacta_ and other publications in 5 languages.

> Address: Apimondia
> Roma, Corso Vittorio Emanuele 101
> Italy

The Bee Research Association has organized a committee on each continent. The American Committee is chaired by Prof. G.F. Townsend, Department of Environmental Biology, University of Guelph, Guelph, Ontario. B.R.A. published _Apicultural Abstracts_, _Bee World_, and _Journal of Apicultural Research_, and provides many services to its members: computer literature searches, book and reprint loans from its library, publications, translations, etc.

> Address: Bee Research Association
> Hill House, Chalfont St. Peter
> Garrards Cross, Bucks.
> England

National. V.G. Milum compiled a history of the national organizations in the U.S. from 1860 to 1964 [342]. The last one he lists, American Beekeeping Federation, publishes a _Newsletter_ now in its 28th volume. In 1969, the National Producer's Organization was also formed. The U.S.D.A. Agriculture Handbook No. 335 [F3] (p. 131-33) gives additional information about various trade groups. Since 1940, the Canadian Beekeepers' Council, composed of representatives from provincial associations and cooperatives has functioned as a coordinating voice for Canadian beekeepers.

Regional, provincial or state, and local. Many of these organizations publish newsletters and reports which do not usually appear in published indexes. Annual reports and proceedings of annual meetings have variously been published by official government agencies: Boards of Horticulture or Entomology, Commissions on Agriculture, Divisions and Departments of Agriculture, etc. There is no list of local organizations available but information can usually be obtained from provincial, state, and county extension agents, apiary inspectors or apiarists, and bee journals carry notices of meetings and conventions.

BEEKEEPING JOURNALS

There are approximately 100 journals devoted exclusively to apiculture being published in the world at the present time. Among them the American Bee Journal, Gleanings in Bee Culture and the Canadian Bee Journal were established in 1861, 1873, and 1885 respectively. A French Canadian journal, L'Abeille et l'Erable, began in 1924. Commercial enterprises have contributed significantly to the dissemination of information on the history of the craft, technical advances and scientific investigations through sponsoring publication of beekeeping journals. More than 110 such journals have been in existence for periods varying from one month to several years before merging with another journal or dropping out of sight. A.I. Root launched Gleanings in Bee Culture which included his price lists as supplements. The American Bee Journal was acquired by Dadant & Sons in 1912. In addition, a considerable number of books and pamphlets that appear in this bibliography were published by these two manufacturing concerns.

The Union List of Serials gives the libraries where these and other journals can be found, and the extent of each library's holdings. The following refer specifically to beekeeping journals:

Bee Research Association. nd List of journals (abbreviated) in English known to contain publications on bees, beekeeping, etc. 11p

Bee Research Association. 1968 World list of beekeeping journals and other serial publications received by the library. Bee World 49(3): 121-40 [Repr L7]

Crane, E. 1955 World list of current beekeeping journals, London, Bee Res. Ass. 12p

Frykholm, L. 1950 Förtechning över Bitidskrifter [List of bee journals]. Bitidningen 49(1): 8-11; (2): 39-45

Frykholm, L. 1951 [List of bee journals.] Suppl. Bitidningen 50(5): 139-42

Frykholm, L. 1954 Förtechning över Bitidskrifter. Uppsala, Kungl. Lantbrukshögskolans Bibliotek. 22p mimeo

Hitz, V.E. & Hawes, I.L. 1930 [F64] p. 167-206

Jacobs, K. 1950 List of serials currently received in the Library of the United States Department of Agriculture.

Miller, C.C. 1897 Defunct bee-journals. Glean. Bee Cult. 25(1): 48-49; (4): 113; (5): 157; (8): 291

Milum, V.G. 1954 Information on beekeeping journals. 12p mimeo

[Wilson, H.F.] 1930 A catalog list of American bee journals including Canada and the United States with a list of the numbers and volumes in the Miller Memorial Library. 15p mimeo Repr. in Rep. of the [Iowa] State Apiarist for 1930 p. 91-100; Rep. of Iowa State Hort. Soc. 65 p. 462-71

TRADE CATALOGS

L.B. Romaine published A Guide to American Trade Catalogs 1744-1900
[New York, R.R. Bowker 422p (1960)], but included only 9 concerns
dealing in bee supplies. The following list, also incomplete, is from
files of catalogs in the libraries visited and from advertising pages of
books and journals. The Miller Collection (Univ. of Wis.), although un-
cataloged, is the most extensive. Dates refer to the year the catalog
was issued or cited, and an underscored date [1880] indicates the year the
company is known to have been organized or ceased operation. The
companies known to be currently operating are marked with an asterisk [*].

See F10 for list of dealers.

*C.W. Aeppler Co	Oconomowoc, Wis.	1931-1942
Aliso Apiary	El Toro, Calif.	
H. Alley	Wenham, Mass.	1875
C.E. Atchley [formerly Jenny Atchley Co]	Beeville, Tex.	1904
Babcock Honey Co	Columbia, S.C.	
Wm. Bamber	Mt. Pleasant, Mich.	
Bee-Maid Mfg. Co	Beloit, Wis.	1961
Blanke Mfg. & Supply Co [formerly Blanke & Hauke Supply Co]	St. Louis, Mo.	1913
Bockhauss Industries	Pagosa Springs, Colo.	
Brew Mfg. Co	Puyallup, Wash.	1940
W.W. Cary & Son	Lyonville, Mass.	1860-1910
Central Supply House	Topeka, Kans.	1918
*W.A. Chrysler & Son	Chatham, Ont.	1880-
Colorado Honey Prod. Ass.	Denver, Colo.	1915
J.H.M. Cook	New York, N.Y.	3rd 1889,1903
C.W. Costellow	Waterboro, Maine	1885
Coomb & Sons	Guilford, Vt.	
*Dadant & Sons	Hamilton, Ill.	1863-
Wm. Daniels	Stoystown, Pa.	1889
Hilas D. Davis	Bradford, Vt.	1885
*Diamond Match Co	Chico, Calif.	1917-
Gus. Dittmer Co	Augusta. Wis.	1907-1914
John Doll & Son	Minneapolis, Minn.	1905
G.M. Doolittle	[Borodino, N.Y.]	1883
Drake & Smith [successors to A.E. Manum]	Bristol, Vt.	1885
Dixie Wood Works	Belton, S.C.	1955
B.O. Everett	Toledo, Ohio	1878
W.T. Falconer Mfg. Co	Jamestown, N.Y.	1885-1930
F.H. Farmer	Boston, Mass.	1905
Griggs Bros. Co	Toledo, Ohio	1909
A.H. Hale Co	Hapeville, Ga.	1953
The Ham & Nott Co. Ltd	Brantford, Ont.	1907
James Heddon	Dowagiac, Mich.	1887,1890
A.G. Hill	Kendallville, Ind.	1888
*S.P. Hodgson & Sons	New Westminister, B.C.	1926-
Chas. E. Hopper & Co	Toronto, Ont.	1916
Houck & Peet	Canajoharie, N.Y.	1882
*Hubbard Apiaries	Onsted, Mich.	1940-

Interstate Box & Mfg. Co	Hudson, Wis.	1902
B.L. Johnson & Co. Inc.	Roaring River, N.C.	1925
D.A. Jones	Beeton, Ont.	1885
*F.W. Jones & Sons, Ltd	Bedford, Que.	
*Walter T. Kelley Co	Clarkson, Ky.	1933-
[Gulf Coast Bee Co	Houma, La.	1925-1933]
Kehm Bros. Apiaries	Grand Island, Nebr.	1963
Kretchmer Mfg. Co	Council Bluffs, Iowa	1864-1919
[E. Kretchmer	Coburg, Iowa	1864]
*Leahy Mfg. Co	Higginsville, Mo.	1884-
G.B. Lewis Co [purchased by Dadant	Watertown, Wis.	1889-
& Sons, 1957]		
Lewis & Parks	Watertown, Wis.	1874-1889
*C.T. Loewen & Sons Ltd	Steinback, Man.	1932-
August Lotz Co	Boyd, Wis.	1904-
[Aug. Lotz & Son	Cadott, Wis.	1904]
A.E. Manum	Bristol, Vt.	1879
Marshfield Mfg. Co	Marshfield, Wis.	1896-1932
Myers Craft Mfg. Co	Burgaw, N.C.	1959-1967
Minnesota Bee Supply Co	Buffalo, Minn.	1927-1942
	[Minneapolis, Minn.	1908]
Charles Mondeng Co	Minneapolis, Minn.	1897-1932
Moody & Isham	Waybridge, Vt.	1885
R.R. Murphy	Garden Plains, Ill.	1878
Chas. F. Muth & Son	Cincinnati, Ohio	1858-
°[Fred W. Muth Co.]		
National Supply Co	Elgin, Ill.	
J.H. Nellis	Canajoharie, N.Y.	1878
E.R. Newcomb	Pleasant Valley, N.Y.	1889
Thomas G. Newman & Son	Chicago, Ill.	1899
Earl M. Nichols [successor to W.W.	Lyonville, Mass.	1910-1913
Cary & Son]		
Joseph Nysewander	Des Moines, Iowa	1899-1907
J. Oatman & Sons	Dundee, Ill.	1879
*Melford Olson Honey Co	Minneapolis, Minn.	1970-
Page & Lyon Mfg. Co	New London, Wis.	1900-1909
The Penn Co	Penn, Miss.	1919
E.L. Pratt [pseud Swarthmore]	Marlboro, Mass.	1889
	[Beverly, Mass.	1891]
Walter S. Pouder	Indianapolis, Ind.	1890
W.H. Putnam [publ The Rural Bee	River Falls, Wis.	1887
Keeper]		
F.J. Rettig & Sons	Wabash, Ind.	1925
M. Richardson	Port Colborne, Ont.	1879
River Falls Bee-Keepers Supply Co	River Falls, Wis.	
[Successor to W.H. Putnam]		
*A.I. Root Co	Medina, Ohio	1869-
J.W. Rouse & Co	Mexico, Mo.	
Ruddy Mfg. Co Ltd	Brantford, Ont.	1928
A.H. Rusch & Son Co	Reedsville, Wis.	1894-1953
J.C. & H.P. Sayles	Hartford, Wis.	1879
R.H. Schmidt Co	Sheboygan, Wis.	1903
C.M. Scott & Co	Indianapolis, Ind.	1896-1903
J.M. Shuck	Des Moines, Iowa	1885
*C.L. Stonecypher	Homerville, Ga.	1954-
The Stover Apiaries	Mayhew, Miss.	1932-1941

I.J. Stringham	Glen Cove, L.I., N.Y.	1898-1913
		22nd
Superior Honey Co	Ogden, Utah	1928-1942
G.L. Tinker	New Philadelphia, Ohio	1889
S. Valentine & Son	Hagerstown, Maine	1885
J.A. Van Deusen	Sprout Brook, N.Y.	
C.H.W. Weber	Cincinnati, Ohio	1904-1907
Geo. T. Wheeler	Mexico, N.Y.	1875
White Mfg. Co	Blossom, Tex.	1900-1905
Williams Bro. Mfg. Co	Portland, Oreg.	1961
*A.G. Woodman Co	Grand Rapids, Mich.	1880-
Geo. W. York & Co	Chicago, Ill.	1903

BIBLIOGRAPHIC SOURCES

Various efforts have been made to organize the extensive world
literature in apiculture such as that begun by Dr. Manger of Germany in
1920 [Bee World 4(11): 216 (1923)]. E.F. Phillips of the U.S. Department
of Agriculture initiated a bibliography which is kept up-to-date and can
be consulted in the Bee Culture Branch Library in Beltsville, Md. [Glean.
Bee Cult. 38(2): 46 (1910)]. Subject headings for O.W. Park's bibliography of 16,000 entries were published in Report of the [Iowa] State
Apiarist for 1954 (p. 88-103).

The 10,000 scientific publications through 1933 listed by D.M.T.
Morland while he was in charge of the Bee Department at the Rothamsted
Experimental Station in England form the nucleus of the bibliography now
maintained by the Library of the Bee Research Association. Approximately
5,000 references were added in 1949 to bring it up-to-date and since 1950,
a portion of the current research and technical articles have been listed
in Apicultural Abstracts.

Indexes to periodical articles.

Agricultural Index. 1916-1964
Amer Ass Econ Entomol Index to the literature of American Economic
    Entomology 1905-59.
Bibliography of Agriculture 1942-1969.
Campbell, D.J. & Henderson, G.P. 1962 The Bee World Index to Vol 1-30
    1919-1949. London, B.R.A. xx,119p
Eckert, J.E. & Shaw, F.R. 1960 [157] p. 473-517
Hitz, V.E. & Hawes, I.L. 1930 [F64] p. 69-82
Report of the [Iowa] state apiarist 1912-54. [See 1926 p. 77-82 for
    contents 1912-22]
Smith, R.C. 1952 rev Guide to the literature of the zoological sciences.
    Minneapolis, Burgess Publ Co 121p
USDA Bibliography of the more important contributions to American
    Economic Entomology 1889-1904
USDA 1932, 1935 Index to publications.

Library collections. The contents of apicultural collections
throughout the world that have been published are listed in Bee World
35(6): 115 (1954); 38(3): 74 (1957). A leaflet is available from the
Bee Research Association concerning use of their library materials
[Repr Bee World 50(1): 21-22 (1969)]. The American Branch of the B.R.A.
Library is located in the Department of Environmental Biology (Apiculture), University of Guelph, Guelph, Ontario c/o Prof. G.F. Townsend.
The unpublished list of the Drory Bibliothek (1903) of the University of
Berlin is available in the U.S. National Agricultural Library and was
incorporated into the Bee Branch Library Bibliography. A photostatic
copy made by the B.R.A. is in the Mann Library, Cornell University. A
description and history of the collections of bee literature in the
National Agricultural Library and university libraries (by J.S. Merrill)
is included in the USDA Agr Handbook No. 335 [F3].

The following citations pertain to U.S. libraries:

Boardman, E. 1966 Bee library collection. Univ Calif. Apiaries,
    Davis 2p
Hitz, V.E. & Hawes, I.L. 1930 [F64]
Int Ass Agr Librarians & Documentalists 1962 U.S.A. National Agricul-
    tural Library Centenary issue 1862-1962. Quarterly Bull 7(2/3):
    97-244
Jager, F. 1929 [245]
Langstroth, L.L. [1895] [285]
Phillips, E.F. [1925?] The Cornell beekeeping library.
Richter, M.C. nd [455]
Walker, H.J.O. 1929 Descriptive catalogue of a library of bee-books
    collected and offered for sale. 144p [Purchased by S.L. Odagard for
    the Miller Memorial Libr]
[Wilson, H.F.] 1925 The Dr. Charles C. Miller Memorial Apicultural
    Library. Wis. Beekeeping 2(6): 57-87
W[orks] P[rogress] A[dministration] 1936 A list of the publications on
    apiculture contained in the Dr. Charles C. Miller Memorial Apiculture
    Library. Madison, Wis., Univ Wis., Coll Agr Libr 283p mimeo

    Theses. A list of theses in apiculture was published in Bee World
50(3): 105-10 (1969) and is also available as a separate leaflet (L8)
from the Bee Research Association.

Dissertation Abstracts. 1938 Ann Arbor, Mich., Univ Microfilms
Doctoral dissertations accepted by American universities 1934--. N.Y.,
    H.W. Wilson Co

    Museums. University departments of apiculture usually have small
collections of equipment as do some dealers and packers. The only
extensive one known to us is the Hewitt Collection [223] at Litchfield,
Conn.

    Films and filmstrips.

Abbott, C.P. & Butler, C.G. 1952 The scientist and the cine camera.
    London, Central Ass Bee-Keepers 7p
Bee Res. Ass. Bee photography group. Bee World 36: 127-28 (1955)
    [B.R.A. library maintains a collection of photographs]
Bee Res. Ass. World list of beekeeping films. Bee World 32: 20-21
    (1951); 51: 6 (1970)
Educational Film Catalog
A directory of bee films. Glean. Bee Cult. 93: 219-28 (1965)
Motion Picture Review Digest. 1936-- N.Y., H.W. Wilson Co
Scientific Film Review. London, International Sci Film Ass [quarterly]
USDA 1948 Slide-films of the U.S. Department of Agriculture. Misc
    Pub No. 655 22p [includes listing of motion picture films]
USDA [Catalog of films with details for loan or purchase. Motion
    Picture Serv, Office of Inform
USDA [Price list of available filmstrips and slides.] Photography
    Div, Office of Inform

Abbott, E.T. (1847-   ) 7
Abbott, M. 8
Abrams, G.J. (1902-1965) Md 1, 2
Adair, D.L. (-1904) 9-12
Adams, C.D. Wis 5
Adams, C.F. 13
Adams, R.L. F20; Calif 1
Adams, T.J. 14
Adie, A. Ont 5, 21
Adrian, M. 579
Affleck, T. 15
Agr Adjustment Admin F21
Aldrich, C.C. 16
Alex, A.H. Tex 9
Alexander, E.W. (1815-1908) 17
Alexander, W.P. NY 1
Alford, F.C. Colo 1
Allen, G. Tenn 3
Allen, J. 95
Allen, T.R. 18
Alley, H. (1835-1908) 19-23
Allinger, H.W. Calif 9
Amer Bee J 612
Amer Honey Inst 24-35, 48, 193,
    548
Amer Honey Prod League 36, 89
Amer Sunday-school Union 111
Amer Tract Soc 42
Amos, J.M. Del 1
Anderson, E.D. F22
Anderson, E.J. Pa 1-6, 8
Anderson, L.D. Calif 2
Argo, V.N. Ohio 1
Arnett, C.M. Fla 1
Arnold, D.J. 37
Arnold, L.E. Fla 2
Arnott, J.H. Sask 2
L'Arrivee, J.C.M. Cl, 10
Atchley, J. (1857-1927) 38
Atherton, G. 39
Atkins, E.L., Jr. Calif 2
Atkins, E.W. 40
Atwater, H.W. F65
Atwood, F.J. Nebr 5
Atwood, G.G. NY 6
Avebury, J. Lubbock (1834-1913)
    41
B., de 42
Baerg, W.J. Ark 1
Bain, F.W. 43
Bain, H.M. F23
Baker, E.L. Conn 1
Baldwin, C.H. Ind 3

Baldwin, E.G. 44; Ind 1, 2
Baldwin, L. 45
Ballantine, W. 46, 47
Barnard, H.E. (-1947) 48
Barnes, G. Ark 1
Barnes, J. 49
Barnes, W. 49
Barth, W. 136, 512
Bartholomew, C.E. Tenn 1
Bartman, M. 50
Barton, B.S. (1766-1815) 51
Baskin, C.C. Ala 1
Bauer, F.W. Calif 3
Beard, D.F. Ohio 2
Beck, B.F. (1868-1942) 52, 53
Beecher, H.W. Vt 1
Bee-keepers' Review 5
Beekeepers State of Ariz Ariz 2
Beland, H. Que 2
Belknap, J. (1744-1798) 54
Benedict, A. 55
Benson, P., Sr. 335
Bently, G.M. (-1955) Tenn 2
Benton, F. (1852-1919) 56, 57,
    383; F24, 25
Benton, R. (1884-   ) Mont 1
Berlepsch, A. 58
Berliner, J.J. & Staff 59
Bess, H.A. Hawaii 2
Bessey, C.E. Nebr 1
Bessonet, E.C. 60
Bevan, E. (1770-1866) 61
Bieberdorf, G.A. Okla 2
Bigelow, E.F. (1860-1938) 62
Biggle, J. 313
Bill, A.C. 63
Birkett, T.E. Mo 1
Bishop, R.K. F137
Bisson, C.S. (1891-1940) F26, 141
Black, J.H. Calif 13
Blake, G.H. Ala 1
Bland, S.E. Sask 2-4
Boardman, E.
Boggs, N. Colo 2
Bohart, G.E. F27; Utah 1
Bonsels, W. 64
Bossé, L. Que 3
Bostwick, E.P. F28
Boswell, P. 65
Bourne, A.I. Mass 1
Boy Scouts of America 420
Bradshaw, D.B. 66
Brady, G.D. Ont 1

Braun, E. (1906-1957)   C2-6
Bray, C.H.   Ky 1
Brice, B.A.   F28
Brite, C.   67
Broadman, J.   68
Brose, C.M.   Colo 3
Brown, A.C.   69, 524
Brown, H.   70
Brown, J.P.H.   (1831-1909)   71
Brown, W.   72
Browne, C.A.   (1870-1947)   F11,
    29, 30
Bruckisch, W.   156
Bryan, A.H.   F31
Buchanan, J.M.   Tenn 3
Burdette, R.F.   Md 2
Bur Cooperative Ass   NY 2
Bur Human Nutrition & Home Econ
    F32,33
Burgess, T.W.   73
Burgin, C.J.   Tex 2
Burke, P.W.   Ont 1-5, 18, 21, 23
Burnside, C.E.   (1895-1949)   F34-
    36
Burrier, A.S.   Oreg 1
Burroughs, J.   (1837-1921)   74,
    75
Burt, M.E.   (-1918)   76
Busch, W.   77
Buschbauer, H.   229
Busdicker, H.B.   Wash 7
Butler, F.   (1766-1843)   78
Butler, G.D.   Ariz 3
Butler, L.I.   Wash 7
Buttel-Reepen, H.v.   79
Byrn, M.L.   80
Cale, G.H.   (1890-1965)   81, 82,
    233; Md 4
Cale, G.H., Jr.   (1919-   )
    83, 84
Calif Honey Advis Bd   85, 517-
    519
Calkins, R.E.   F114
Cantin, C.   Que 4
Carl, F.L.   Mo 2
Carpenter, C.G.   86
Carr, E.G.   F37; NJ 1-7, 20;
    NC 1
Casteel, D.B.   F38, 39
Caster, A.B.   Ariz 5
Centennial Bee-Keepers Ass   87
Central Wash Beekeepers Ass
    Wash 1
Chadwick, P.C.   88
Chambers, M.D.   89
Chase, A.W.   90

Chercuitte, U.   C7
Childers, L.F.   (1878-1961)   Mo 6
Childs, A.M.   Minn 1
Christ, J.L.   547
Claffey, J.B.   F40
Clark, E.H.   91
Clark, J.B.   (1869-   )   92
Clarke, W.F.   (-1902)   93, 94
Clarke, W.W., Jr.   Pa 1, 7, 8
Clements, F.E.   F41
Clericus   165
Clute, O.   (1839-1902)   95
Cockerham, K.L.   La 1
Coggshall, W.L.   NY 3
Cohn, D.J.   96
Colberg, W.J.   NDak 5
Coleman, G.A.   97; Calif 4
Coleman, M.L.   98
Coley, H.W.   Conn 2
Collings, ?   99
Colvin, R.   100
Comstock, A.B.   (1854-1930)   101
Conn Beekeepers Ass   102
Conservative Bee Keeper   110
Cook, A.J.   (1842-1916)   87, 103,
    383; F42; Mich 2-4
Cook, E.H.   104
Cooper, J.F.   (1789-1851)   105
Corkins, C.L.   Wyo 1-5, 11
Cornell Apis Club   106
Corner, J.   BC 1
Cory, E.N.   Md 5
Cotton, C.B.   107, 109
Cotton, L.E.   108, 109
Cotton, W.C.   110
Country Curate   169
Crandall, L.B.   Conn 3-7
Crane, J.E.   (1840-1930)   Vt 2
Crews, A.   Wash 2
Crompton, J.   282
Cross, J.H.   111
Crowder, D.E.   112
Cunningham, W.F.   113
Curtis, G. DeC.   (1870-   )   114
Cushman, S.   (-1933)   RI 1-3
Cutting, J.A.   115
Dadant & Sons   116, 284
Dadant, C.   (1817-1902)   117, 284,
    413
Dadant, C.P.   (1851-1938)   117-
    122, 284, 374; Ill 1
Dadant, J.C.   (1911-1954)   120
Dadant, M.G.   120, 123, 340
Dalton, S.   142
Daly, H.V.   Calif 5
Danzenbaker, F.   (1837-   )   124

Richmond, R.G.   Colo 9-11
Richter, M.C.   455; Calif 16
Ridgway, R.L.   Tex 2
Riebel, L.   456
Ritchie, M.   457
Roat, J.   341
Roberts, J.F.   BC 5
Roberts, W.C.   F73
Robertson, C. (1858-   )   458
Robertson, D.R.   Man 6-10
Robinson, S.   459
Rock, J.P.   Vt 3
Rodriguez, D.A.   F123
Rokahl, M.   Wyo 11
Roney, J.N.   Ariz 7, 8
Rood, R.N.   460
Root, A.I., Co   461-493
Root, A.I. (1839-1923)   494-496
Root, E.R. (1862-1953)   124, 339,
    419, 494, 496-508, 512; Mass
    15; Ohio 8
Root, H.H.   17, 124, 509 (1873-1972)
Root, L.C. (1840-1928)   442
Rosen, E.   510
Rouse, J.W. (1852-1937)   511
Rowe, H.G. (1870-1934)   512
Rowe, I.B.   Minn 10
Rowell, J.O.   Va 1
Rowland, M.J.   Ont 19
Rude, C.S.   Okla 4; Tex 14
Ruehe, H.A.   Ill 13
Ruffin, W.A.   Ala 2
Russell, F.   513
Sackett, W.G.   Colo 12
Sanborn, C.E. (-1946) Okla 5-
    10; Tex 12
Sandburg, C. (1878-   )   514
Sanmann, F.F.   Ill 13
Santerre, A.   515
Sauriol, C.   516
Savery, A. 575
Schaefer, C.W.   F53, 124
Schafer, M.   517-519
Schamu, C.G.   520
Scheibner, R.A.   Ky 8
Schmitz, G.H.   F125
Schofield, A.N.   521
Scholl, E.E.   Tex 3, 12
Scholl, L.H. (1880-1956)   Tex 13
Scullen, H.A.   Ohio 9; Oreg 1,
    5-7, 9
Sechrist, E.L. (1873-1953)   522-
    525; F126-130
Secor, E. (1841-1919)   341, 526,
    527
Sein, F., Jr.   F131

Shafer, G.D.   Mich 25
Sharp, D.L. (1870-1930)   528
Shaw, F.R. (1908-   )   157, 418,
    596; Mass 2, 13, 16, 17; Okla 4
Sheldon, C.   529
Sheppard, W.J. (-1935)   530,
    531; BC 4, 5
Sherman, F.   NC 2
Sherman, L.   Wyo 11
Sherwood, S.   F31
Shuttlesworth, D. (1907-   )   532
Singh, S.   NY 24
Sippel, O.A.   Mont 2
Sisson, L.S.   533
Sisson, D.   439
Skinner, D.   534
Skinner, T.C.   Fla 8
Sladen, F.W.L. (1876-1921)   C23-26
Slater, L.G.   535
Slocum, B.A.   Wash 3, 8-15
Smedley, D.   53
Smith, C.D.   Mich 26
Smith, C.H.   536
Smith, D.L.   Man 10, 11
Smith, E.C.   537
Smith, F.G.   538
Smith, H.   539
Smith, J. (1871-1953)   540-543
Smith, J.V.C. (1800-1879)   158, 544
Smith, M.V.   NY 25; Ont 16-18, 24,
    25
Smith, W.D.   545
Snodgrass, R.E. (1875-1962)   546;
    F132, 133
Sorenson, C.J.   Utah 5, 6
Souder, D.   547
Southerland, F.L.   F28
Southern Calif Honey Exhibit Comm
    548
Spangler, R.L.   F128
Sparks, L.M.   SC 3
Spectator   549
Spence, W.   274
Stanger, W.   Calif 10, 13, 17-23
Stephen, W.A.   NC 4-6; Ohio 3, 10
Stephen, W.P.   Oreg 8
Stevens, R.O.   NC 7
Stewart, C.   F16
Stiles, C.F.   Okla 4, 11
Stockley, C.   550
Stockwell, G.A. (1847-1912)   551
Strong, R.G.   La 3
Strutt, E.   552
Stuart, F.S.   553
Sturges, A.M. (1879-1934)   554
Sturtevant, A.P. (1890-1960)   F35,

Wiestling, J.S.   598
Wilder, J.J. (1873-1950)   599-601;
    Fla 10, 13
Wildman, D.   562
Wiley, H.W. (1844-1930)   6; F162
Wilhelm, M.   602
Wilkes, A.H.   603
Wilkin, R. (1829-1901)   604
Williams, G.W.   605
Willson, R.B.   NY 26
Wilson, C.A.   Miss 1
Wilson, H.F. (1883-1959)   Oreg 9;
    Wis 9-14; p. 88; p. 93
Wilson, W.T.   Colo 7, 13
Wis State Beekeepers' Ass   606
Wolf, C.W. (-1866)   607
Wolff, W.H.   NH 3
Wood, J.G. (1827-1889)   608
Woodman, A.G., Co.   609
Woodrow, A.W.   F163-165; NY 29
Woods, B.   NY 14
Woodward, F.W.   244
Woodward, G.E.   244
Working, D.W.   610
W[orks] P[rogress] A[dmin]   50;
    p. 93
Wright, A.T.   611
Wright, W.D. (1851-1933)   NY 28
Yates, A.W. (1864-1943)   Conn 8
Yeatman, F.W.   F161
York, G.W. (1862-1937)   341, 612
Yost, C.O. (-1931)   Ind 14
Young, W.J.   F30
Yrisarri, E.   568
Zinn, H.J.   613
Van Zwaluwenburg, R.H.   F166

www.ingramcontent.com/pod-product-compliance
Lightning Source LLC
Chambersburg PA
CBHW051338200326
41519CB00026B/7470